THE LIVELY MEMBRANES

THE LIVELY MEMBRANES

R. N. Robertson

CAMBRIDGE UNIVERSITY PRESS

Cambridge

London New York New Rochelle

Melbourne Sydney

10-28-85

Published by the Press Syndicate of the University of Cambridge
The Pitt Building, Trumpington Street, Cambridge CB2 1RP
32 East 57th Street, New York, NY 10022, USA
296 Beaconsfield Parade, Middle Park, Melbourne 3206, Australia

© Cambridge University Press 1983

First published 1983

Printed in Great Britain by the University Press, Cambridge

Library of Congress Catalogue card number: 82–22011

British Library cataloguing in publication data
Robertson, R. N.
The lively membranes.
1. Membranes (Biology)
I. Title
574.87′5   QH601

ISBN 0 521 23747 5 hard covers
ISBN 0 521 28202 0 paperback

U P

CONTENTS

PREFACE

Library shelves are filling with books on membranes. Why, then, should I presume to write another? There are several reasons.

I want to give the reader a picture of how lively these remarkable structures are at the molecular level. To visualise what is happening needs a moving picture in the mind, difficult to convey in writing and best done with an animated film which, restricted to a two-dimensional view, would still fall short of conveying their dynamic three-dimensional character. Here, with diagrams and text, I attempt to stimulate the reader's imagination.

My particular approach in visualising membrane molecules in motion follows a tradition in chemical modelling which goes back to Kekulé with his visions, on top of a London bus sometime before 1858, of the way atoms danced and joined together in molecules, a tradition very productive of concepts. This is an essay, not a treatise; it minimises formal mathematical analysis, not because that is unimportant, indeed it is essential since imaginative pictures must be checked with quantitative data. Here, with brief but appropriate evidence, I am attempting to present a stimulating and exciting account of membranes and some of their activities.

Since so many of the reactions of living systems occur in or on membranes, every biologist needs to know about their basic properties – knowledge which is now, perhaps, as essential to biologists as the periodic table has been to chemists. I am, therefore, writing a book which everyone, including those who have done biology and chemistry at secondary school only, should be able to read. Inevitably this leads to some simplifications. The difficulties of writing such a book in another field were well expressed 70 years ago by J. C. Stobart, author of *The Glory that was Greece* (1911):

The writer who attempts (to present a panorama of the whole territory from an individual point of view) will, of course, be inviting criticism at a thousand points. He is compelled to deal in large generalizations and to tread upon innumerable toes with every step he takes. Every fact he chronicles is the subject of a monograph, every opinion he hazards may run counter to somebody's life-work. He will often have to neglect the latest theory and sometimes he is unaware of the latest discovery.

I crave the patience of my specialist colleagues who do not need the book anyway, unless they use it to improve or correct a concept which I have stated imperfectly or too speculatively.

Above all I hope this will be a readable book, with a text that is enjoyed and illustrations that are appreciated. Selected references are given, especially to review articles and books, for those who wish to follow the subject further. For many people, not steeped in biochemistry, it will be helpful to refer from time to time to an appropriate textbook, for example Stryer's book (1981), clear and easily read, or Lehninger's (1975), more detailed and comprehensive but also easily understood.

Rutherford Robertson

Sydney, 1982

Lehninger, A. L. (1975). *Biochemistry*, 2nd edn. New York: Worth Publishers Inc.
Stobart, J. C. (1911). *The Glory that was Greece*. London: Sidgwick & Jackson.
Stryer, L. (1981). *Biochemistry*, 2nd edn. San Francisco: W. H. Freeman & Co.

ACKNOWLEDGEMENTS

'It's a very ancient saying but a true and honest thought,
That, if you become a teacher, by your pupils you'll be taught.'
The King and I. Rodgers & Hammerstein.

I am grateful to the many students and colleagues who have contributed to my scientific education, but especially to my teacher, Professor G. E. Briggs, F.R.S.

My thanks to those who read parts of the manuscript for me: Dr John Charnock, Professor Peter Gage, Dr Jacob Israelachvili, Dr Tony Larkum and Dr Ivan Ryrie. Special gratitude to Adele Post, my former graduate assistant, who helped in many ways.

For electron micrographs, I am indebted to Dr J. M. Bain (then of CSIRO Food Research Laboratories), Dr A. D. Blest (Research School of Biological Sciences, Australian National University) and Dr S. K. Fisher (Biology Department, University of California, Santa Barbara).

The diagrams, which are a central feature of this book, were done by Mrs Sandra Smith, Canberra.

Permission for reproduction was granted by: Academic Press (Fig. 1.2a, Fig. 1.6, Table 11.1); *Science*, American Association for the Advancement of Science (Fig. 1.5); *Quarterly Review of Biophysics*, Cambridge University Press (Table 3.1); W. H. Freeman and Company, *Biochemistry*, 2nd ed. © 1981 (Fig. 7.1); *Journal of Cellular and Comparative Physiology* (Fig. 1.4); Royal Society of London (Fig. 1.3), and by the following authors: A. D. Blest, S. K. Fisher, R. Henderson, J. Israelachvili, J. Oró, S. J. Singer and L. Stryer. I am grateful to Dr A. D. Bangham, F.R.S., for permission to describe a variation of his experiment (pp. 117–8).

Above all, I am indebted to my wife without whom this book would not have been written and with whom it has been a continuing and enjoyable partnership.

ix

1

Lively properties – themes and variations

Very many different kinds of membranes occur throughout the living world. They are responsible for a wide variety of activities. The most widely known membranes are those which, on the surfaces of cells, separate the inside from the outside so that the composition of the solution within the cell can differ from that outside. For many years biologists believed that this behaviour, as barrier to movement of substances in solution, was the principal function of membranes. But many membranes do much more than simply maintain a difference of composition. In this chapter, I shall outline some examples of the many functions of membranes.

Where differences in composition of solutions on either side of the membrane occur, the membrane itself is often responsible for creating these differences. Various kinds of *pumps*, sited in the membranes themselves, control the composition on the two sides, e.g. high potassium on one side and high sodium on the other. Such membranes, which have pumps, maintain the differences between the cell cytoplasm and the interior of organelles (mitochondria, plastids, Golgi bodies, nuclei, endoplasmic reticulum and others). Electron micrographs of animal and plant cells showing their membranes are seen in Fig. 1.1(*a*) and (*b*); Fig. 1.1(*c*) and (*d*) show diagrammatic representations of the types of membranes which can occur in animal cells (*c*), and in plant cells (*d*).

Another important function is illustrated by the membranes of the thylakoids in the chloroplasts of green plants and in the surface membranes of the photosynthetic bacteria, which trap light energy and transform it into other kinds of energy, notably electrical and chemical energy. The membranes in nerve cells have the vital function of transmitting electrical signals by successive depolarisations along their lengths. Eyes have

1

Fig. 1.1. (*a*) Membranes in animal (chicken liver) cells and (*b*) in plant (mung bean stem) cells, as shown by the electron microscope. Each shows part of three adjoining cells. M, mitochondria; N, nucleus; NM, nuclear membrane; CM, cytoplasmic membrane; ER, endoplasmic reticulum; GB, Golgi body; CW, cell wall; P, plastid; SG, starch grain; PL, plasmalemma; T, tonoplast; V, vacuole. (Micrographs: J. M. Bain.) (*c*) Types of membranes which can occur in animal cells and (*d*) in plant cells. Each line represents a membrane. Nu, nucleolus; Ve, vesicle; L, lysosome; G, grana; Th, thylakoid; RER, rough endoplasmic reticulum; SER, smooth endoplasmic reticulum; NA, nerve axon; Sy, synapse; SyVe, synaptic vesicles; Exc, exocytosis; Enc, endocytosis.

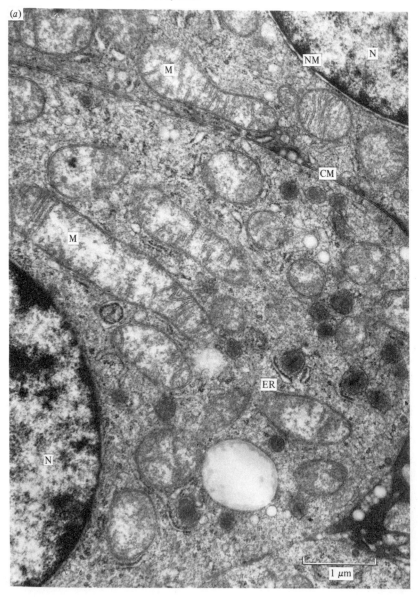

3

Fig. 1.1. (b)

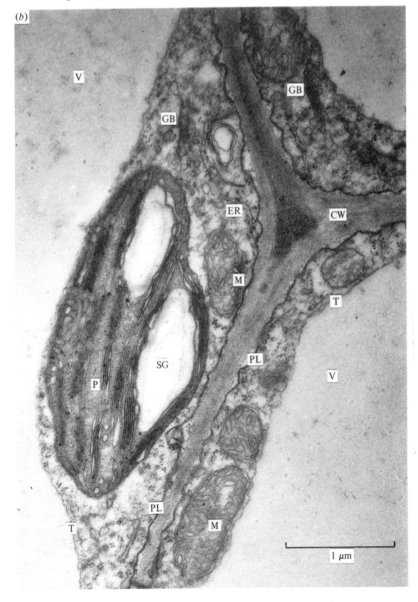

4

Fig. 1.1. (c)

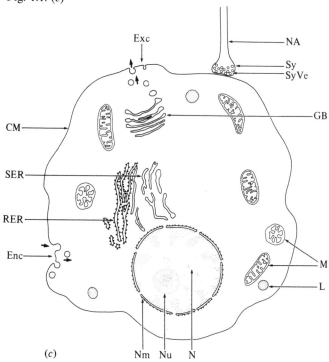

(c)

Fig. 1.1. (d)

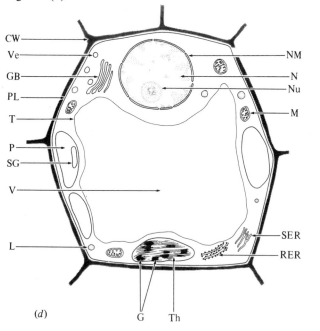

(d)

membranes which trap light and give rise to electrical nerve impulses (Fig. 1.2).

The membranes of mitochondria and of respiring bacterial cells are responsible for the conversion of the energy of oxidation-reduction reactions into the formation of adenosine triphosphate (ATP), the

Fig. 1.2. Specialised cells for light absorption in vertebrate eyes are known as retinal rods and have characteristic stacks of membranes: (a) an electron micrograph of a section of the outer segment of a retinal rod in the eye of a ground squirrel, showing the bounding membrane of the cell, CM, and the stack of enclosed membrane discs, DS. Note the abundant mitochondria, M, lower in the cell (micrograph: D. H. Anderson & S. K. Fisher, 1976); (b) section through a retinal rod from a cat's eye shows the double membranes of the discs within the cell membrane (micrograph: S. K. Fisher).

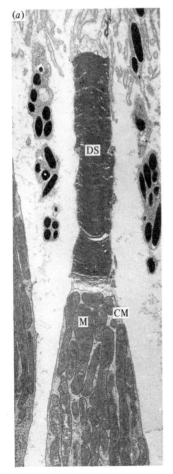
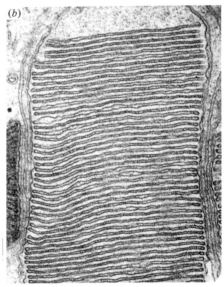

commonest energy currency of the cell. Some membranes of the endo-plasmic reticulum are associated with ribosomes and their protein synthesis, and others are the sites of lipid synthesis, not only for maintaining the membrane structure but also for other purposes. Cell-surface membranes contain the systems which, on recognising substances which are antigens, transmit the message so that antibodies are synthesised inside the cells. Not least important in all membranes is the way in which parts of the molecules at the centre of the membrane provide a non-aqueous environment where, in conjunction with the appropriate portions of membrane enzymes, chemical reactions may occur in ways which would be impossible in aqueous solution.

Some membranes take substances into cells, some push them out, sometimes via the formation of vesicles which may bud off from, or snap on to the inside of the cell-surface membrane. Membranes are very dynamic structures, being continually formed and maintained, or

Fig. 1.3. Membranes in the outer segment of the eye of the spider *Dinopus*, which sees its prey in the dark, increase rapidly at night and are broken down by day: (*a*) membranes as they appear in daytime; (*b*) membranes greatly increased and organised into microvilli within 60 min at dusk (micrographs: A. D. Blest, 1978).

(*a*)

(*b*)

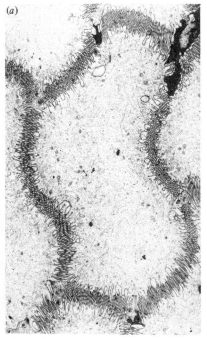
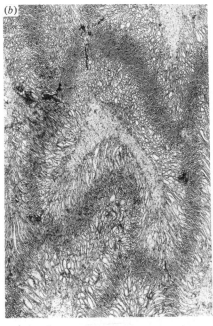

dismantled and their components metabolised, according to the cell's requirements at the time (Fig. 1.3). Furthermore their adaptability, once associated with the origins of cellular life, has allowed the evolution of many different functions in different organs and organisms.

Despite the very wide diversity of their activities, illustrated by these examples, the basic structures and properties of all membranes are, at first sight, surprisingly similar, and the rest of this chapter discusses their common properties and some variations in their functions.

So thin

Membranes are very thin and are not visible under a light microscope. It is true that in cells or organelles it is possible to see the optical effect at a curved interface where the membrane will be, but we do not see the membrane. The beautiful diagrams of membranes drawn after the living material has been fixed, sectioned and observed with the electron microscope, are sometimes misleading when applied as preconceived ideas to interpretation of light microscope fields. A few years ago I was visiting the Harvard Biological Laboratories where my friends were bemoaning the way in which freshmen students, brought up on the splendid diagrams of biology textbooks, would look at a cell under a student microscope and ask the instructor to show them the endoplasmic reticulum. I repeated this story when I later visited Cambridge University; there, a colleague said 'Oh, our students are better than that; they sit at the students' microscopes and draw the endoplasmic reticulum!' If you think of the cell as the size of an average lecture room, the membranes on such a scale would be only about 4 mm thick but, with the cell at its ordinary size, not optically resolvable.

Membranes are basically only two molecules thick. About 45 years ago I arranged a discussion, in an Australian university, on the nature of protoplasm. Such was the state of understanding then, that a professor of biochemistry told a physical chemist that he had read of the possibility that membranes were only two molecules thick. 'Could that be so?' he asked in a voice of incredulity. The physical chemist replied that he would be very surprised if they were more than two molecules thick, a reply that was greeted with good-humoured disbelief.

Understanding should have progressed beyond that point, even in remote Australia, because that was approximately 10 years after the classic experiment of Gorter & Grendel (1925). They took some of the membranes which were the empty envelopes of erythrocytes, extracted the lipids, dissolved the lipids in petroleum ether and then allowed them to spread

8

at an air-water interface. The monomolecular film so formed turned out to have an area twice that of the erythrocytes from which the lipid molecules had been extracted, consistent with the possibility that the natural membrane could have been a layer two lipid molecules thick. Their estimate of thickness was 4 nm. This basic idea was developed further but turned out to be substantially correct and the term *bilayer* is now widely used as a description of the basic structural framework of living membranes.

The first detailed discussion of the probable properties of bilayers was due to Danielli & Davson. In 1935 they published a diagram to reconcile the presence of proteins with the lipids of the membrane. In 1936 Danielli prepared a membrane around a drop of salt solution containing a little protein, by allowing it to fall through the interface between an oil and a protein solution. At first we thought that the proteins might be spread on each side of the membrane at the interfaces between the water and the membrane lipids. The model of Danielli & Davson is shown in Fig. 1.4.

Fig. 1.4. The early model of the membrane proposed by Danielli & Davson (1935): (A) the layers of lipid molecules with their polar heads towards water (B) on each side of the membrane; proteins (C) were pictured as being spread on the polar heads of the lipids.

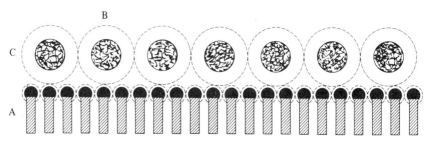

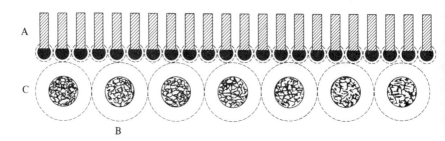

They discussed this concept and reviewed much of the evidence for various features of membrane structure and function in their important book, *The Permeability of Natural Membranes* (Davson & Danielli, 1943). Subsequently, additional information was building up about the roles of the proteins which, with the lipids, are in different proportions in different membranes and it gradually became clear that there are proteins on each side, partially in the water and partially buried in the membrane. This realisation culminated in the suggestion of Singer & Nicolson (1972) that the membrane is basically composed of two layers of fluid lipids in which protein molecules are buried to different extents. This model, which became known as the 'fluid mosaic model', is illustrated in Fig. 1.5.

The essential properties of membranes depend on the bilayer consisting of orientated lipid molecules with their *hydrocarbon chains*, or fatty tails, meeting in the centre and their *hydrophilic*, or water-loving, *head groups* in the water on each side. The centre of the membrane is, then, a thin zone

Fig. 1.5. The fluid mosaic model of Singer & Nicolson (1972): the predominantly fluid lipid molecules (A) form the bilayer; proteins (B) may span (B_1) or rest in (B_2) the bilayer. (Redrawn from Singer & Nicolson, copyright 1972 by the American Association for the Advancement of Science.)

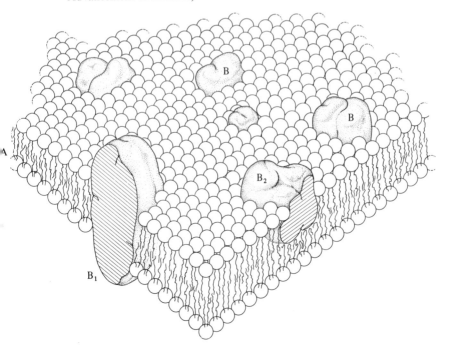

where the hydrocarbon chains make it fat-loving, or *lipophilic*, and water-hating or *hydrophobic*. But membranes are very varied in structure and no single model describes them all exactly, especially as a single membrane is not necessarily homogeneous. Some of the simpler possible variations in different parts of a membrane (Israelachvili, 1978) are shown diagrammatically in Fig. 1.6. Needless to say, membranes' functional diversity depends on their structural differences as well as on their different protein compositions.

Fig. 1.6. Variations in the structures within a membrane; the membrane consisting of a bilayer of lipid molecules (A), is folded (A_1); it contains some crystalline patches of lipid molecules (A_2); some proteins (B_1) span the membrane with hydrophilic portions on both sides; others (B_2) are set in the membrane to different depths; one protein (B_3) is shown trapped between the two folds of the membrane; the structure of a lipid pore is shown at (C). (Israelachvili, 1978.)

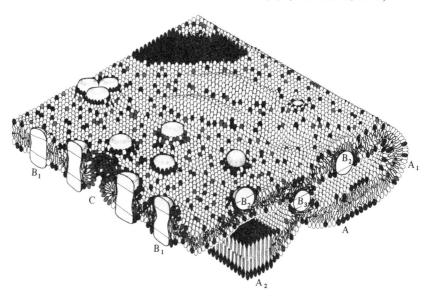

Stable but flexible

These bimolecular membranes or bilayers have properties which, at first sight, may appear somewhat paradoxical. They retain their basic sheet-like stability although they may alter their shapes substantially, e.g. by bending, always provided they are not rapidly stretched too far, up to about 4%. Thus, they are stable and flexible but inelastic. Every biologist has probably observed an animal cell bursting when it is transferred from an isotonic solution to one of low osmotic pressure. The same thing happens to naked plant protoplasts or to plant cells (like those of apple fruits) in which the cellulose cell walls are so thin that the cells swell in water, stretching the walls until the membranes puncture and the cell contents are lost. Under a phase contrast microscope it is easy to observe mitochondria swelling in a hypotonic solution until they suddenly disappear when, as the membranes burst, the optical effect is lost.

Strong barriers to water-soluble molecules and ions

Membranes are highly impermeable to water-soluble molecules and especially impermeable to ions. This property enables them to play one of their important roles in the control mechanisms of the cell. The diffusive movement of substances in a solution is governed by two principal

Table 1.1. *Diffusion coefficients in dilute aqueous solutions*

Substance	Molecular weight	Temperature (°C)	$D \times 10^5$ (cm^2/s)	Reference
H$_2$	2	21	5.2	a
O$_2$	32	18	1.98	a
HCl	36.5	19	2.5	a
CO$_2$	44	20	1.77	a
NaCl	58.5	20	1.39	a
Urea	60	20	1.18	a
KCl	74.5	20	1.68	a
Glycerol	92	20	0.83	a
Lactose	342	20	0.43	a
Raffinose	504	20	0.36	a
Myoglobin	17 500	20	0.11	b
Lactoglobulin	37 900	20	0.07	b
Haemoglobin	68 000	20	0.06	b
Edestin (seed globulin)	309 000	20	0.04	b

a *International Critical Tables* (1929), vol. 5, p. 63. New York: McGraw-Hill.
b T. Svedberg (1938). *Ind. Eng. Chem., Anal. Ed.*, **10**, 113.

related functions; first, the size of the diffusing particle and second, the solubility, that is the amount of the dissolved substances likely to be in a given volume of solvent at any time. The diffusion of different substances in different liquids can be compared by using the *diffusivity* (diffusion coefficient) which is the amount diffusing in unit time, across unit area for unit distance with unit concentration drop and is expressed as cm^2/s. This unit looks puzzling on first acquaintance; the amounts in mol, volumes in cm^3, areas in cm^2 and distances in cm cancel to cm^2. Some of the differences in diffusivity in water are striking and are illustrated in Table 1.1. The diffusivity increases approximately logarithmically with decreasing molecular weight.

A membrane, unlike a solvent in which a solute is diffusing, is not homogeneous in composition across its width and we cannot use diffusivity to apply to these bilayers. We therefore use the *permeability coefficient*, which is the amount of solute diffusing across unit area of membrane in unit time for unit concentration drop and, in the units already mentioned, becomes cm/s. Suppose a salt such as potassium chloride were diffusing across a short distance of water (e.g. 5 nm) and came to equilibrium with a half-time of 1 ms. If we replaced that 5 nm of water with a bilayer in which the permeability coefficient for potassium chloride was only about 10^{-9} times that in water, a reasonable figure for membranes, the time to reach equilibrium would be about 10^9 ms or about 280 hours (12 days!). Membranes are very real barriers to diffusion. Furthermore, as we shall see, the permeability is very different for different substances. Thus, water may have a permeability coefficient in a membrane of 10^{-2} cm/s; in the same membrane the permeability coefficient of sodium might be 10^{-12} cm/s, about ten thousand million times slower than water! Water in a *Chara* cell has a permeability coefficient of about 2×10^{-2} cm/s, whereas the quite small molecule of urea has a permeability coefficient of about 1×10^{-6} cm/s, i.e. about 10000 times slower in diffusion. Membranes with these high impermeabilities provide the cell with structures to control the substances in solution on each side of the membranes. However, despite their low permeabilities when intact, they are under the control of the cell and their properties can be modified so that if a barrier becomes unnecessary the cell can render the membrane permeable or destroy it altogether.

Very effective insulators

High resistance to current, or effective insulation, is not unconnected with the impermeability to ions, both effects arising from the hydrophobic interior of membranes. In contrast to their behaviour in water, charged particles or ions in any numbers cannot move through these membranes, thereby reducing ion conduction of current to exceedingly low values. Further, the inner portion of the bilayer consists of a predominantly hydrocarbon environment which, since hydrocarbons are the best organic insulators, thus prevents passage of current by electron flow. Within this insulating lipid medium in some membranes, there are some restricted protein channels specifically for conduction of electrons, protons or ions, but these are not a general feature of membranes. Where they have been developed, these channels are very important and necessary for controlling the electrical potential difference and the composition on each side. It is the high insulation that makes it possible to have channels for specialised conduction.

Rapid water penetration

It is a common biological observation that, compared with the slowness of the movement of dissolved substances, water moves rapidly through living membranes; for example, plasmolysed plant cells can be seen to regain their turgor without loss of a dissolved substance from inside or penetration of a dissolved substance from outside. The rapid movement seems to some people, at first sight, to be surprising. Why can water molecules penetrate these membranes when dissolved substances cannot? Very simply water, a *small* uncharged molecule, is not completely insoluble in the lipophilic region of the membrane so there are always some water molecules there but they are few in number and isolated from each other. If the concentration of water is greater on one side of the membrane than on the other, water will move through until equilibrium is reached on the two sides. In contrast, as we have seen, the substances which dissolve in water are larger, often charged and so are less likely to penetrate. The speed of diffusion of water through a membrane, compared to that of ions, can be great – as we saw for sodium, thousands of million times faster.

Specific substances may move across

If there is an effective barrier to most substances, special pathways for particular molecules or ions required by the cell or organelle may be developed in the membranes in the form of specific channels. The principle is illustrated by the antibiotic *gramicidin* which is not used to increase the permeability of the cell which makes and secretes it but to increase the permeability of membranes in other organisms such as bacteria, with disastrous results to a would-be invading bacterium. When gramicidin enters the potential competitor, it forms a tube spanning the membrane and allows the passage of K^+ and H^+ through the channel in the centre of its molecule thereby destroying the ion balance in the enemy and killing it. Sometimes specific transport takes the form of a carrier soluble in the membrane, which picks up a substance on one side, diffuses through the membrane and, when it meets water on the other side, loses the substance to the water. These carriers are often specialised proteins or parts of proteins but, as we shall see, not necessarily. It is a small theoretical step from the idea of a specific carrier to the idea of a pump in which a carrier is linked to an energy-providing system so that a substance can be transferred across the membrane against a concentration gradient or, in the case of ions, against an electrochemical potential gradient.

Maintaining electrical potential differences

Many membranes have the property of building an electrical potential difference between the two sides and maintaining that difference. Potential differences are very important, for example in redistributing ions on the two sides, in being a mechanism for the storage of readily available energy and, in nerve and some other cells, for the mechanism which transmits the electrical impulse from one end of the cell to the other.

Light trapping

Some membranes, e.g. those of the plant chloroplast thylakoids, of photosynthetic bacteria and of the eyes of animals, trap light energy by means of pigments attached to proteins fixed in the bilayer. One of the characteristics of such membranes is the speed with which a potential difference is established between one side of the membrane and the other. After light is absorbed by the thylakoid, the potential difference is set up in about 10^{-8} s. In such membranes, the potential difference is maintained until the energy is used in secondary effects such as the photosynthetic reactions in thylakoids or reactions leading to nerve impulses in eyes.

Lively molecular movement

In this short account of the similarities and differences between living membranes, it has not been possible to convey the intense activity and movement taking place at the molecular level. Despite the membrane's overall stability and low permeability, the lipid molecules are much less like the pickets in a fence than they are like a surging line of riot police, selectively repelling the penetration of rioters but allowing the entry of law-abiding citizens. Sometimes the penetration is repulsed at the surface by riot shields and batons, sometimes by closing shoulders so the whole line resists invasion. The charged head groups of the lipid molecules resemble the riot shields and batons, repelling ions electrically before they actually penetrate the membrane line. At other times, the tighter packing of the membrane molecules can be likened to the linked-arm, shoulder-to-shoulder resistance of the police and a would-be penetrating molecule is repulsed like an over-zealous rioter. The membrane molecules, like the police, are never still; they change position in the line, they may call up reinforcements, sometimes an individual molecule may reach out of the line like a policeman grabbing someone and pulling him back in, or they may have the equivalent of an armoured car to pick up a law-abiding citizen on one side and carry him through to the other – a carrier molecule such as a protein.

Environment for synthesis

Despite their extreme thinness, membranes of living cells offer three different environments which can affect the behaviour of molecules. First, there are the watery solutions on each side in which the concentration of ions changes the closer they are to the surface charges on the membrane. Then there is the environment among the head groups, themselves often ionised and therefore 'choosey' about which ions or charged parts of molecules can share it with them. Finally, there is the central region of the bilayer which is virtually an oily hydrocarbon in which molecules or parts of molecules which are lipophilic, will dissolve. These differences allow for small molecules, or side groups of large molecules, to be transferred from one kind of environment to another, and so affect their properties. The change from water to oil surroundings is just as real as when an organic chemist arranges one kind of reaction with a molecule in an aqueous phase and another in a non-polar solvent. It is significant that the later stages of synthesis of the membrane lipids take place in the membranes themselves. So, too, does the synthesis of most proteins which are membrane-bound and have lipophilic side chains. On the other hand, some proteins are synthesised in the membrane and extruded from it to become water soluble molecules with hydrophilic groups toward their outsides.

In this chapter, I have discussed examples of the widely differing functions of living membranes but also attempted to show that they have similar basic properties – thin, not more than two lipid molecules in thickness with associated proteins, stable but flexible, strong barriers to water soluble molecules and ions, but permeable to water. These properties enable some to convert light energy into other forms, especially into electric potential differences.

In the subsequent chapters I shall discuss the types of molecules which make dynamic activities possible, the forces which hold them together as membranes, the reactions which they can carry out and finally, how important these molecules and their membranes must have been in the origins and evolution of living organisms as we know them.

Suggested reading

Finean, J. B., Coleman, R. & Michell, R. H. (1978). *Membranes and their Cellular Functions*, 2nd edn. Oxford: Blackwell Scientific Publications (Chap. 1).

Quinn, P. J. (1976). *The Molecular Biology of Cell Membranes*. London and Basingstoke: The Macmillan Press Ltd. (pp. 1–15).

Saier, M. H., Jr. & Stiles, C. D. (1975). *Molecular Dynamics in Biological Membranes*. New York: Springer-Verlag.

2

Composition – special molecules

The lively behaviour of the membranes in cells can be understood only if we know the chemistry of the constituent molecules, the way they are shaped and fitted together in the different kinds of membranes and the way the molecules and parts of molecules move in relation to particular functions. In this chapter we shall look at the nature of the special molecules.

The basic structure of membranes, two molecules in thickness, depends on the properties of lipid molecules and of membrane-associated proteins. These types of molecules, which can be partly hydrophilic and partly

Fig. 2.1. Water spreads and wets a hydrophilic surface like glass, but does not spread on a hydrophobic surface like paraffin.

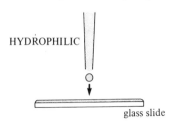

HYDROPHILIC

glass slide

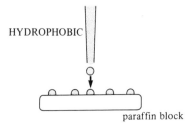

HYDROPHOBIC

paraffin block

lipophilic are said to be *amphipathic*. Our understanding of this double property is helped when we remember the properties of hydrophilic and hydrophobic substances and the molecular explanation of their particular behaviours. If a drop of water falls on a surface which it 'wets', i.e. on a hydrophilic surface like glass, the water immediately spreads thinly. By contrast, if the drop falls on a hydrophobic surface like a solid paraffin which does not wet, the water does not spread but remains confined in drops (Fig. 2.1).

These differences can be pictured as due to the different attractive forces between the molecules of the substances concerned. Surfaces which are wetted by water consist of molecules which can form bonds (particularly hydrogen bonds) with water similar to those between water molecules themselves. The molecules of non-wetted surfaces are not capable of forming such strong bonds with water. Whereas wettable substances have molecules capable of hydrogen bond formation, the molecules of non-wettable substances are held together by the weaker *hydrophobic forces* also known as the *van der Waals forces*. This property is associated with the way they are squeezed out of an aqueous environment. If, however, a molecule is amphipathic, one part will be attracted to water and the other part to a non-aqueous, hydrophobic region. Such molecules are also known as *surface active* because they accumulate at an interface between water and a non-aqueous environment, aligning so that their hydrophilic ends or sides face the water and their hydrophobic ends or sides are in the non-aqueous phase, e.g. air or oil or hydrocarbons. The understanding of the properties of these substances began with Benjamin Franklin's famous experiment on spreading amphipathic oils on the pond of Clapham Common in 1765. In a letter to William Brownrigg he described the spreading oil 'becoming so thin as to produce prismatic Colours for a considerable Space and beyond them so much thinner as to be invisible except in its effect of smoothing waters at a much greater distance'. We now know that his observation that the oil became invisible would be where it was less than 2.5 nm in thickness so light would not be reflected. More than a century was to pass before the theoretical consequences of this observation were to be realised. Using a technique developed by Fraulein Pockels in 1891, Lord Rayleigh showed, some eight years later, that the molecules of monomolecular films on a water surface – for that is what Franklin's oil had become – could be closely packed together by compression in the surface and become a solid film. In 1913 Sir William Hardy pointed out that molecules which form a layer one molecule thick on water, characteristically each have a hydrophilic and a hydrophobic part, an

Fig. 2.2. Molecules at an air–water interface have their hydrophilic
parts in the water and their hydrophobic parts in air: (*a*) a few
molecules remain separated as in a gas; (*b*) more molecules packed
closer together behave like a liquid film; (*c*) when packed tightly
together intermolecular forces act to hold the molecules together in a
solid film.

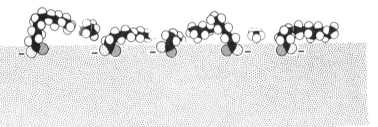

a)

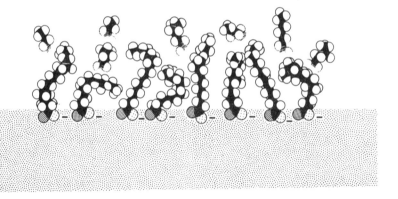

b)

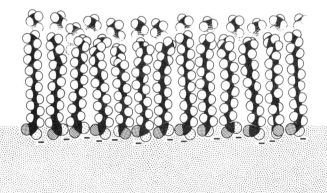

c)

hypothesis elegantly confirmed by the classic work of Langmuir (1917) (Fig. 2.2). Writing in 1937, N. K. Adam said:

> I well remember how some of the orthodox chemists of that time (about 1920) ridiculed the pictures drawn by Langmuir and Harkins, of the molecules of long chain compounds standing on end in surfaces, with their water-soluble ends in the water and their hydrocarbon ends away from it; they said that their constitutional formulae were never meant to be taken as tangibly as that!

Monomolecular layers can be packed in three different forms at an interface, depending on their concentration, on their lateral compression in the surface and on the temperature. If there are few uncompressed molecules at the interface they will move freely, independently of each other, colliding and separating (Fig. 2.2a). As more molecules come into the same area they will more nearly be in a liquid condition, i.e. they are frequently in contact but the forces of attraction between them are insufficient to overcome their kinetic movement so they are diffusing past each other (Fig. 2.2b). But as more molecules are present at the interface and, especially if the temperature is lowered, the attractive forces of the hydrophilic parts between adjacent molecules and the attraction of the hydrophobic parts between adjacent molecules may be stronger than the kinetic movement, and the monolayer can become a rigid crystalline solid structure (Fig. 2.2c). These observations on monomolecular layers established the principles which made it possible to understand what might happen in the bimolecular layers which constitute the living membranes. For example, as we shall see, many molecules in a cellular bilayer behave as though they are in a liquid state but sometimes they pass into the solid state. Furthermore, some molecules can align with their neighbours so that some parts of the molecules behave like a solid while other parts continue to behave like liquids. This peculiar property, very important in the control of diffusion within a membrane, will be discussed in Chapter 4.

Fat-like molecules – lipids
The chemical nature of the hydrophobic portions

Common experience shows that the hydrocarbons (which have repeating chains consisting only of units of —HCH—) are among the most water-repellent substances known – proverbially, oil and water do not mix. It is not surprising that a large proportion of the lipids which make up the bilayers of living membranes have hydrocarbon chains as their hydrophobic portions. A substance of this type, a fatty acid (palmitic), is illustrated in

21

Fig. 2.3. The structure is shown both in the conventional chemical notation (2.3*a*) and in the structural model of the atoms (2.3*b*), not only in the dimensions of the space which they fill but also with the bond angles between them. Such a model, though not perfect, gives a fairly accurate idea of the shape of the molecule and the space which its different parts will occupy. It is important to realise that atoms can rotate around each single bond within the limit of the bond angle. Consequently a molecule, within which the atoms are moving all the time, goes into a variety of shapes (Fig. 2.3*c*, *d*, *e*).

The hydrocarbon or acyl chains which occur in the lipid substances common in living membranes, range from those with 12 carbon atoms (lauric acid) to those with 18 carbon atoms (linolenic) with a few above 20. Some are fully saturated (lauric, myristic, palmitic, stearic), some have one double bond (oleic and palmitoleic), some two (linoleic) and some three (linolenic). Space-filling models of examples of these substances are shown in Fig. 2.4.

These molecules, in constant movement between the molecules of any solvent in which they are dissolved, are also continually changing their shapes by rotation about the single bonds and bending and twisting into many different shapes. A double bond introduces a firm constraint because two parts of a molecule, linked by such a bond, cannot rotate about it.

Fig. 2.3. The anion of palmitic acid: (*a*) in conventional notation; (*b*) in space-filling atomic models, with black carbons, white hydrogens and dotted oxygen, double dotted with a double bond; (*c*), (*d*) and (*e*) are the kinds of shapes which the molecule can assume; (*f*) a water molecule for scale.

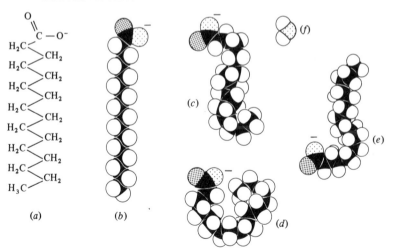

22

Fig. 2.4. Fatty acid anions with different numbers of double bonds: (a) stearic acid – no double bond; (b) oleic acid – one double bond; (c) linoleic acid – two double bonds; (d) linolenic acid – three double bonds.

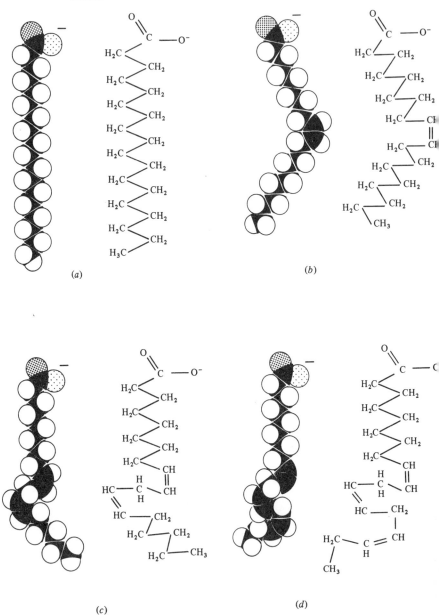

(a)

(b)

(c)

(d)

As we shall see in more detail, the common lipids of membranes usually have *two* hydrocarbon chains. In these double-chained lipids, the fatty acyl chains are usually unbranched, but one of the two is often unsaturated. We shall see that the double chains are essential to the correct solid geometry which allows lipids to form membranes and that the unsaturated chain helps to maintain fluidity in the bilayer. The two lipid chains are not necessarily of the same length. Double-chained lipids are discussed in the next section and examples are shown in Figs. 2.6–2.9.

Certain lipophilic sterols, quite different in chemical structure from the lipids we have been discussing, are also included in membranes. The commonest sterol in membranes is cholesterol which owes its amphipathic character to its hydroxyl group (Fig. 2.5).

The hydrophilic portions

The various compounds with hydrocarbon chains, which occur in the living membranes, have different hydrophilic groups at the end of the chain or chains. The most widely distributed lipids have hydrocarbon chains attached to the phosphate ester of glycerol. The phosphate is on the primary hydroxyl of the glycerol and is usually attached by another ester linkage to a base or to inositol or to glycerol. The fatty acyl chains, usually unbranched but one often unsaturated, are attached by ester links on carbon atoms 1 and 2 (Fig. 2.6). The standard pattern of the phospholipids is shown in Fig. 2.6*h*, but we find a variety, differing not only in the nature of the acyl chains but also in the composition of the hydrophilic groups esterified to the phosphate. The structures of the commonest of these phospholipids are shown by the space-filling atomic models (Fig. 2.7). A particularly interesting phospholipid, cardiolipin, has

Fig. 2.5. (*a*) Cholesterol: the space-filling model shows the tightly packed atoms associated with the ring structures. (*b*) Rotation is possible only about the carbon atoms of the hydrocarbon portion.

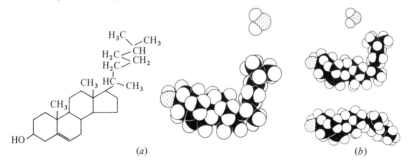

(*a*)   (*b*)

24

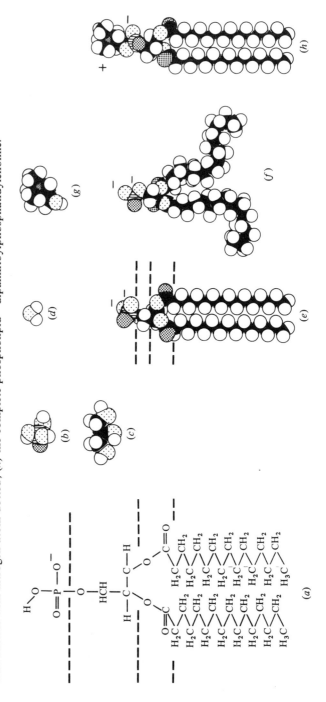

Fig. 2.6. Putting together a phospholipid: (a) dipalmitoylphosphatidic acid showing the ester linkages of the phosphate and of the two palmitic acids to the glycerol; (b) model of phosphate with phosphorus atom hatched; (c) model of glycerol; (d) water; (e) model of dipalmitoylphosphatidic acid; (f) the molecule can take different shapes with rotation about the C—C bonds; (g) the base choline with nitrogen atom dotted; (h) the complete phospholipid – dipalmitoylphosphatidylcholine.

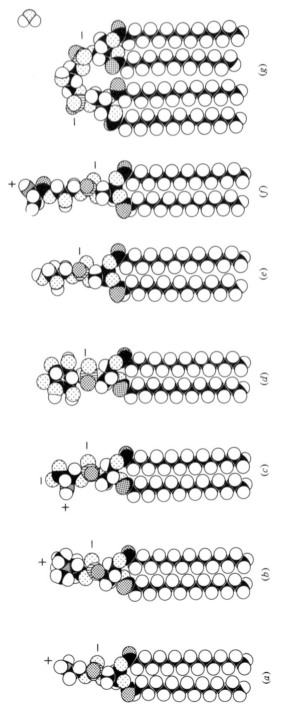

Fig. 2.7. The different polar groups of the phospholipids: (a) phosphatidylethanolamine; (b) phosphatidylcholine; (c) phosphatidylserine; (d) phosphatidylinositol; (e) phosphatidylglycerol; (f) phosphatidyl-O-alanyl glycerol; (g) cardiolipin. All are shown attached to saturated hydrocarbon chains. Some polar groups have positive and negative charges which behave as zwitterions (i.e. they balanced each other's charge within the molecule), others have excess negative charge.

two glycerophosphates esterified to a third glycerol and therefore has two negative charges in the polar group and four acyl chains.

In some phospholipids, known as the sphingolipids, the glycerol is replaced by a long-chain amino alcohol. Thus, sphingomyelin has an additional fatty acid on the amino group of sphingosine as well as a choline linked to a phosphate on the terminal hydroxyl group (Fig. 2.8).

Another group of lipids, the glycolipids, does not have phosphate but the free hydroxyl group of the glycerol or of the amino alcohol may be esterified to a sugar. Galactose and digalactose esters of glycerol with two hydrocarbon chains are the major constituents of the chloroplast membranes of higher plants. Galactose esterified to the free hydroxyl group of the amino alcohol, sphingosine, forms a cerebroside. Neutral or charged amino sugars esterified with sphingosine are known as gangliosides. The galactolipids are particularly important in some systems, as we shall see (Fig. 2.9).

Fig. 2.8. Sphingomyelin: (a) the chemical notation shows the choline group linked to the phosphate which is esterified to the long-chain amino alcohol with a fatty acid on the amino group; (b) the space-filling model shows how this phospholipid has a compact hydrophilic head and hydrophobic hydrocarbon chains.

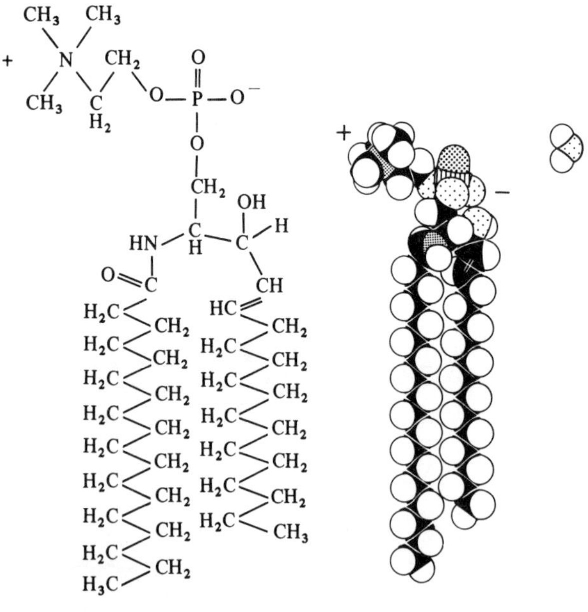

(a)   (b)

With their strongly hydrophilic heads and their lipophilic tails, the lipids will be expected to pack into bilayers, tails towards each other and heads towards the water on each side (Fig. 2.10). Note that the hydrophilic groups of some of the phospholipids lie almost parallel to the surface of the bilayer.

Fig. 2.9. A cerebroside: (a) the notation shows how the hydroxyl group of the long-chain amino alcohol is esterified to galactose; (b) the space-filling model shows the characteristic hydrophobic and hydrophilic portions; (c) note that the longer hydrocarbon chain can fold into various shapes due to the free rotation about C—C bonds and may occupy a greater cross-sectional area than the other chain.

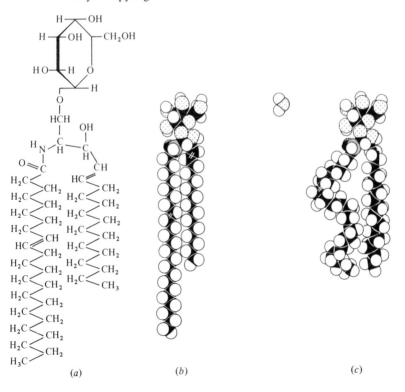

(a)   (b)   (c)

Fig. 2.10. Phospholipid molecules packed into a bilayer with their hydrophilic groups in the water on each side of the film and hydrocarbon chains forming the central lipophilic region. The molecules are not shown as densely packed – hydrophobic chains of adjacent lipids will fill the spaces.

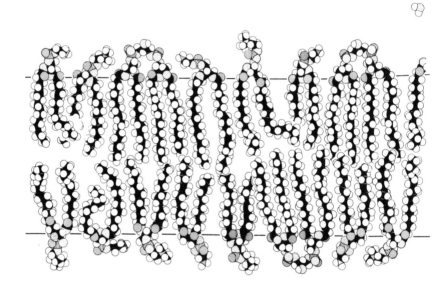

29

Fat-like molecules – non-lipids

Apart from the lipids, some small lipophilic, mostly amphipathic, molecules are quite characteristic of certain membranes and probably exist only in those membranes. Some of these are the prosthetic groups of the membrane-bound enzymes, which will be considered with the proteins, but some, e.g. plastoquinone and ubiquinone, are free in the lipid bilayer where they carry out essential functions in biochemical processes. The carotenoids, e.g. lutein and xanthophyll, which occur in various cells in crystalline and amorphous form, also occur in association with specific proteins in the photosynthetic membranes where they play an important part in trapping light energy. Their properties fit them admirably to be part of the lipophilic phase of the membrane bilayers for they possess an unsaturated hydro-carbon chain. As a model of the structure of carotene shows (Fig. 2.11), they will lie comfortably alongside the hydrocarbons of the lipids. In other words, they will be soluble in the lipophilic phase of the bilayer and also able to have van der Waals interactions with the hydrophobic portions of proteins.

Two very important lipophilic molecules are ubiquinone, which functions in respiratory membranes of mitochondria and bacteria, and plastoquinone, which functions in photosynthetic membranes. Both these molecules are long, consisting predominantly of carbon and hydrogen, but with a group containing oxygens at one end. The space-filling models are shown in Fig. 2.12 and 2.13.

Their 50 carbon-atom chains are made of 10 isoprenoid units. As the

Fig. 2.11. Carotene: this unsaturated hydrocarbon is a hydrophobic pigment which will reside in the lipophilic portion of the bilayer.

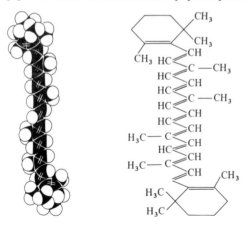

30

Fig. 2.12. Ubiquinone: the highly lipophilic electron–proton carrier of respiration in the inner mitochondrial and bacterial membranes.

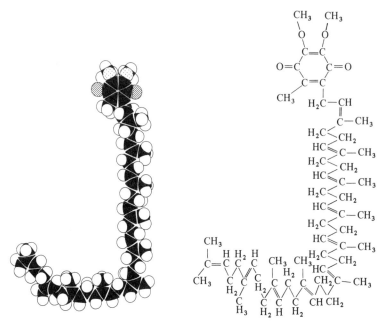

Fig. 2.13. Plastoquinone: the highly lipophilic electron–proton carrier of photosynthesis in the thylakoid membranes of chloroplasts. Note that it differs from ubiquinone only because it has two methyl groups instead of two methoxy groups on the ring.

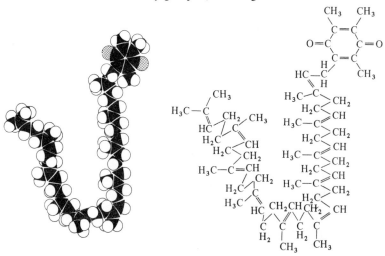

model shows, the isoprenoid tails would be expected to stay, or to dissolve, in the lipophilic part of the bilayer and the heads, containing the oxygens, would tend to find the water on either side of the membrane. The unique importance of these two very lively compounds will be discussed later.

The proteins

The proteins of membranes are responsible for the wide variety of enzyme activities which characterise different membranes. As might be expected, the tightness of binding of the proteins varies with the different types. Some proteins, for example cytochrome c, are soluble in water but function when bound to the membrane surface. Such molecules are called *peripheral* or *extrinsic proteins* because they may come and go with little permanent binding to the membrane; this binding might be achieved by ionic bonds or hydrogen bonds, either with appropriate groups on other membrane proteins, or with the head groups of some lipids. Other peripheral proteins (e.g. the F_1 complex in ATPase) may be held by hydrophobic forces in the lipophilic portions (lipids or proteins) of the bilayer (Fig. 2.14). By contrast, there are *integral* or *intrinsic proteins* which are tightly bound to the membrane, some of which stretch across the bilayer. Such molecules clearly need to accommodate themselves to the different layers of the lipophilic region and of the head groups. Some integral proteins may be on just one side of a membrane, held in position by both the hydrophobic forces of the lipophilic region and the bonds of the head groups.

What sort of structures are to be expected in the membrane-bound proteins to make them compatible with the lipids which, we have seen, make up the basic structure? This question is difficult because of the complexity of protein molecules and the limitations of our knowledge of the structures of particular proteins. The problem is further complicated because many of the proteins of membranes are glycoproteins, i.e. they have amino or neutral sugars covalently bonded, hydrophilic groups because of their ability to form hydrogen bonds. Not many of the membrane proteins have their amino-acid sequences completely known and this makes it difficult to discuss their properties in relation to bonding with adjacent lipids in membranes. It is easy to see that hydrophilic amino-acid residues in the chain would, if they occurred in the right places, be compatible with the lipid head groups and the water on each side of the membrane. Similarly, the neutral or non-polar amino-acid side chains would be compatible with the lipophilic regions of the bilayer but one important difference between proteins and lipids is that proteins are rigid

32

Fig. 2.14. Proteins in a lipid bilayer. Note that the hydrophilic portions of the protein (e.g. with —COOH and —OH groups) are in the water and the hydrophobic portions are in the lipophilic region of the bilayer. Note how the proteins alter the packing of the lipids.

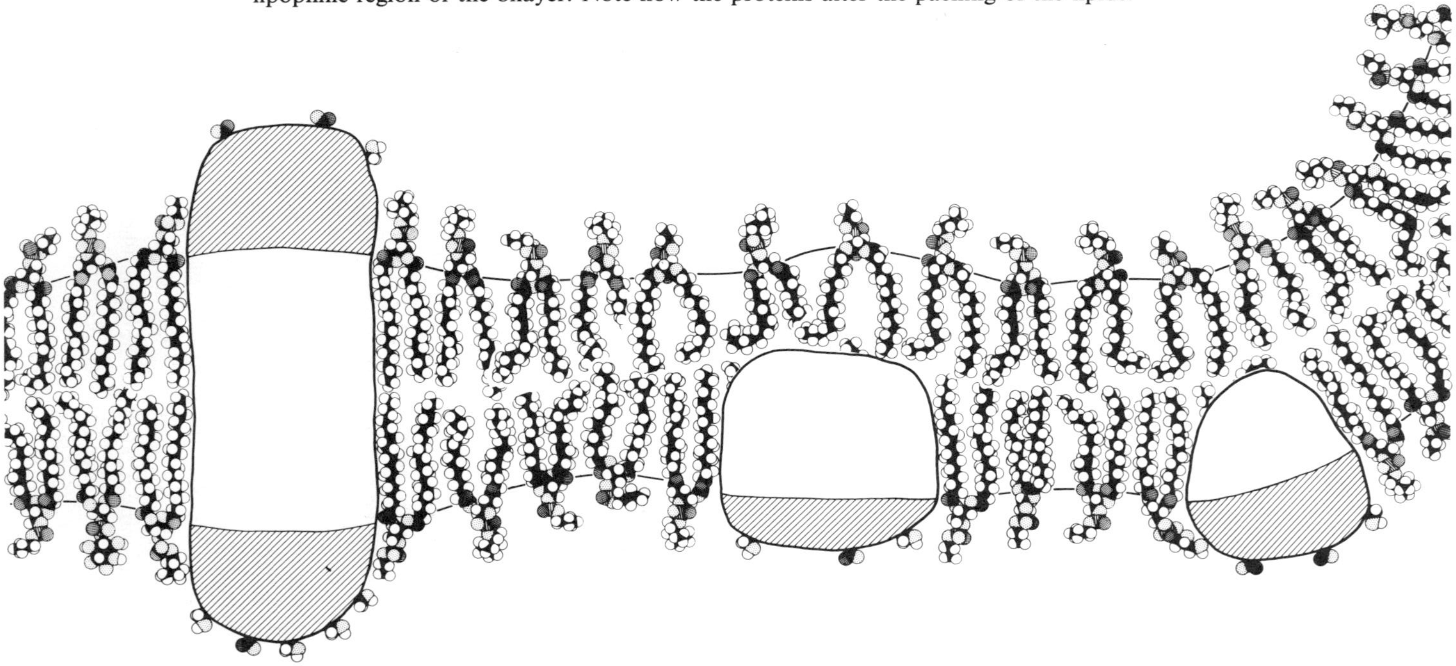

33

structures. As we shall see, though parts of them may change their shapes or conformations, they cannot pack to fill spaces with the fluidity characteristic of the hydrocarbon chains of the lipids.

Because there are so many membrane proteins and because so little is known about their structures in general, it is best to illustrate the principles by discussing four: probably the best understood integral protein is the *bacteriorhodopsin* which is responsible for the light-trapping properties of the bacterium *Halobacterium halobium*; cytochrome *c* is a peripheral protein temporarily combined with the inner mitochondrial membrane; the mammalian cytochrome oxidase which spans the inner mitochondrial membrane; the large protein complex, part integral and part peripheral, of proton-ATPase which is found with relatively minor modifications in mitochondrial, chloroplast and bacterial membranes.

Bacteriorhodopsin

This light-trapping, proton-pumping enzyme, found in *Halobacterium halobium* is very similar to the protein, rhodopsin, found in eyes of all animals. Thanks to very extensive work on bacteriorhodopsin, its sequence and crystalline structure are known and its probable binding within the membrane can be largely understood.

The molecule is a single polypeptide with a molecular weight of 26000,

Fig. 2.15. A diagrammatic representation of the α-helices of bacteriorhodopsin and the way they span the lipid bilayer, as suggested by Henderson (1977). Reproduced, with permission, from the Annual Review of Biophysics and Bioengineering, Volume 6. © 1977 by Annual Reviews Inc.

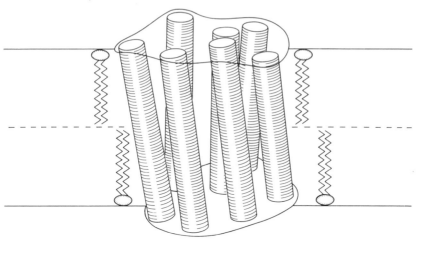

34

purple in colour due to the presence of retinal, the aldehyde of vitamin A. One molecule of this low molecular weight retinal is bound to a particular lysine in the polypeptide chain and we shall return to its properties later. Meanwhile, we are interested in the form of the whole molecule. By various techniques, but especially those of X-ray and electron diffraction, it has been shown that the molecule consists of 7 rods, each about 4 nm long and 1 nm apart (Henderson, 1977). The rods span the lipophilic portion of the membrane, three approximately vertical in the plane of the membrane and

Fig. 2.16. A more detailed representation of the bacteriorhodopsin molecule adapted from Engelman *et al.* (1980) to show the polypeptide chain connecting the α-helices and the positions of the amino-acid residues. All amino acids in the helices are neutral except where shown. Insert (*a*) shows how the atomic model would appear at (*a₁*). The approximate dimension (dashed line) (*b*) of retinal (*b₁*) and its position if attached to lysine 216 are also shown.

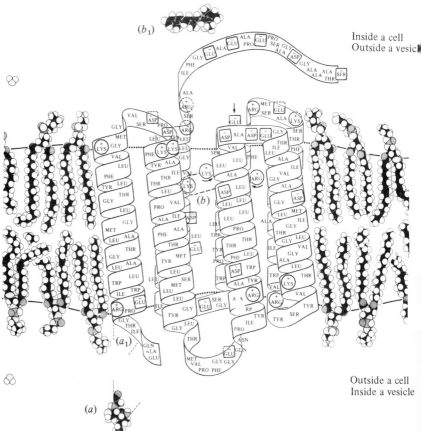

four tilted by 10° to 20° (Fig. 2.15). Each rod is an α-helix of the polypeptide chain and, since the complete sequence of the protein is known (Gerber *et al.*, 1979; Ovchinnikov *et al.*, 1979), it has been possible to show that some of the chain connecting the seven rods is on one side of the membrane and some on the other (Engelman *et al.*, 1980). We see from Fig. 2.16 that most of the amino-acid side chains of the rods are neutral. By contrast, the parts of the protein which are near or protrude into the water on the inside of the membrane contain 19 charged amino-acid residues and the parts which protrude into the water on the outside of the membrane contain six charged groups. Within the rods there are nine charged groups, five negative, four positive, which may be sufficiently close to satisfy each others' charges in the predominantly non-aqueous environment. For our present purposes we note that:

(1) This large molecule, spanning the membrane, with seven α-helices with predominantly hydrophobic side chains can be expected to pack into the lipophilic region, the 4 nm length of hydrophobic surface on each α-helix being about twice the length of a lipid chain;

(2) the charge portions of the protein chain project into, or are close to, the water on each side of the membrane;

(3) apart from some limited movement of side chains, the hydrophobic part of the molecule will be relatively rigid.

Cytochrome c

The molecule of cytochrome *c* (m.w. 12 500) is water soluble, occurring between the inner and outer membranes of mitochondria. It functions by combining temporarily with other membrane-bound components of the electron transport chain (cytochrome reductase and cytochrome oxidase) in the inner mitochondrial membrane. The structure of cytochrome *c* is well understood (Dickerson, 1972). It consists of a single polypeptide chain of 104 residues and a covalently bound haem group. The molecule is roughly spherical with a diameter of 3.4 nm. Though most of the amino-acid side chains are hydrophobic and wrapped around the haem, some which make contact with water are hydrophilic. A ring of positively charged lysine side chains around the exposed haem edge seems to be responsible for attaching this peripheral molecule by ionic bonds to negatively charged groups in the integral membrane molecules. We shall see the importance of this in our discussion of energy transduction. These lysine groups seem to be a regular feature of cytochrome *c* molecules, conserved during evolution, even though cytochromes from different organisms differ in other respects.

Mammalian cytochrome oxidase

An integral protein complex of molecular weight about 200 000 and length 8.3 nm, this oxidase spans the inner mitochondrial membrane. It consists of six or seven different polypeptides which are present in 1:1 stoichiometry of two molecules of haem A and two copper atoms. Though little is known about the way these polypeptides are packed into the structure which makes them compatible with the membrane, three things, important for our present considerations, are clear: first, the characteristic rigid structure of proteins predominates and has great importance in the way the protein is packed and functions in its membrane. Second, the protein has a band, about 4 nm wide, which is hydrophobic and fits the hydrocarbon portion of the bilayer and, third, the surface polypeptides on one end are known to be compatible with binding to the lysines of cytochrome c.

The proton-ATPase

This enzyme complex occurs bound in some membrane in almost every cell. It can function in two ways: it can hydrolyse ATP (adenosine-triphosphate) to ADP (adenosinediphosphate) and P_i (inorganic phosphate) on one side of the membrane while producing protons on the other; alternatively, if supplied with ADP and P_i on one side and with protons on the other, it can synthesise ATP. We shall be examining its mode of action in later chapters.

Here, we are concerned with its complicated protein structure, as far as it is known. Two distinct portions of the proton-ATPase can be distinguished: one, the peripheral protein region, where the catalytic reactions occur, can be seen by suitable electron micrograph techniques as a knob of diameter about 9 nm, which has a molecular weight of about 380 000, protruding from the membrane; the other portion is an integral protein complex in the bilayer (Fig. 2.17). The peripheral knob is known as F_1 and the integral base portion is known as F_0. Many workers believe that the two are connected by a short stalk with proteins different from either those in the knob or in the base portion. The F_1 which is hydrophilic consists of five types of protein subunits, with stoichiometry α_3, β_3, γ_1, δ_1, ϵ_1. These subunits are now well characterised, can be extracted separately and can be put together again to reconstitute the active enzyme (Alfonzo & Racker, 1979; Kagawa et al., 1979). The F_0 portion, whose protein components are not so well characterised, not only anchors the whole enzyme complex in the membrane, but also provides some sort of mechanism for proton movement. Its proteins are closely associated with

Fig. 2.17. A very diagrammatic mitochondrial ATPase: (a) the morphological knob of proteins which are characterised but not sequenced; (b) the membrane-bound proteins; (c) the hypothetical, but not established, channel for protons. We do not know how the various proteins are aligned.

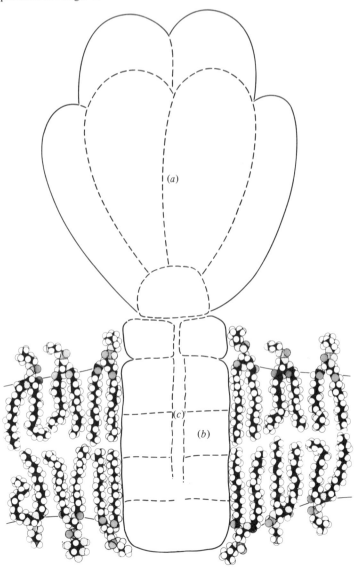

phospholipids which are difficult to remove in extraction procedures; indeed, the detergents used in the extraction often finish up combined with these very hydrophobic proteins. The F_0, including part of the stalk, probably consists of three different proteins which, together, must be associated with phospholipids to carry out their function of translocating protons across the membrane – in one direction when ATP is being synthesised and in the opposite direction when it is being hydrolysed. Activities of this ATPase, from various sources, have been much investigated in bilayers surrounding an aqueous solution forming small vesicles known as *liposomes*. It is clear that the proton pathway through F_0 functions only if it is reconstituted across a bilayer containing phospholipids. The lipids apparently align with, and are tightly held to, the very hydrophobic surfaces of these proteins.

Different membranes, different composition

Since membranes carry out such a wide range of different functions, it is not surprising that they differ widely in composition. As most of the reactions occurring in the membrane are associated with the different protein enzymes, very different proteins are to be expected but a wide range of lipid compositions is also found. The lipids are not merely structural units supporting the proteins but are necessary to the membrane enzymes' functions. Commonly membranes may consist of 50 or more kinds of proteins accompanied by a host of phospholipids and glycolipids with various head groups, numbers of chains, lengths of chains and degrees of unsaturation. Steroids, pigments and other lipophilic molecules will also be packed in as required for different functions.

The ratio of protein to lipid varies greatly in different membranes (Table 2.1). For example, the myelin sheath, a membrane round certain nerve fibres, which functions primarily as an insulator, has a low content of protein (only about 20%, wt/wt). In contrast, the inner mitochondrial

Table 2.1. *Weight per cent lipid and protein in cellular membranes*

Membranes	Lipid	Protein
Myelin	80	20
Chloroplast	50	50
Erythrocyte	40	60
Microsomes	32	68
Mitochondria	26	74

membrane, an energy transducing membrane, has 74% protein and only 26% lipid.

Some of the mitochondrial membrane proteins are buried within the hydrophobic centre of the bilayer unlike the cytochrome oxidase, which traverses the whole bilayer or the cytochrome c, which is superficially and loosely attached. These differences illustrate the variety of properties in proteins to allow them to pack in these different ways – to be discussed more fully in the next chapter.

The amounts of the different lipids vary widely in different membranes. For instance, an ox erythrocyte membrane has cholesterol as 50% of its lipids and less than 5% as phosphatidylcholine. By contrast, the inner membrane of rat liver mitochondria has 41% of phosphatidylcholine and no cholesterol. The reasons for the wide differences in composition are not yet fully understood but, as we shall see, must be closely related to function; for example, the outer and inner membranes of rat liver mitochondria have different percentages of the principal phospholipids (Table 2.2). The hydrocarbon chains of the phospholipids may be of several different lengths and of different numbers of double bonds (Table 2.3). The common phospholipids, phosphatidylcholine and phosphatidylethanolamine, together with an occasional cholesterol, are diagrammatically illustrated

Table 2.2. *Phospholipid content of inner and outer membranes of rat liver mitochondria (%). Adapted from Stoffel & Schiefer (1968)*

	Phosphatidyl-choline	Phosphatidyl-ethanolamine	Cardiolipin	Phosphatidylinositol phosphatidylserine
Inner	41	35	21	2
Outer	49	31	3	17

Table 2.3. *Distribution (per cent) of chain length and double bonds in principal phospholipids of inner membrane of rat liver mitochondria. Adapted from Parkes & Thompson (1970)*

Chain length:no. of double bonds	16:0	18:0	18:1	18:2	20:4
Phosphatidylcholine	18.1	26.7	9.2	35.9	3.1
Phosphatidylethanolamine	5.7	38.6	4.9	18.7	20.2
Cardiolipin	4.6	1.6	6.8	75.4	0.7
Phosphatidylinositol	4.4	57.6	13.6	6.0	10.4

Fig. 2.18. A bilayer of phosphatidylcholine (a), phosphatidylethanolamine (b) and cholesterol (c). The molecules must be pictured as being in constant motion. The dotted lines draw attention to the effective spaces occupied by polar groups, kept apart by charge and bound water molecules.

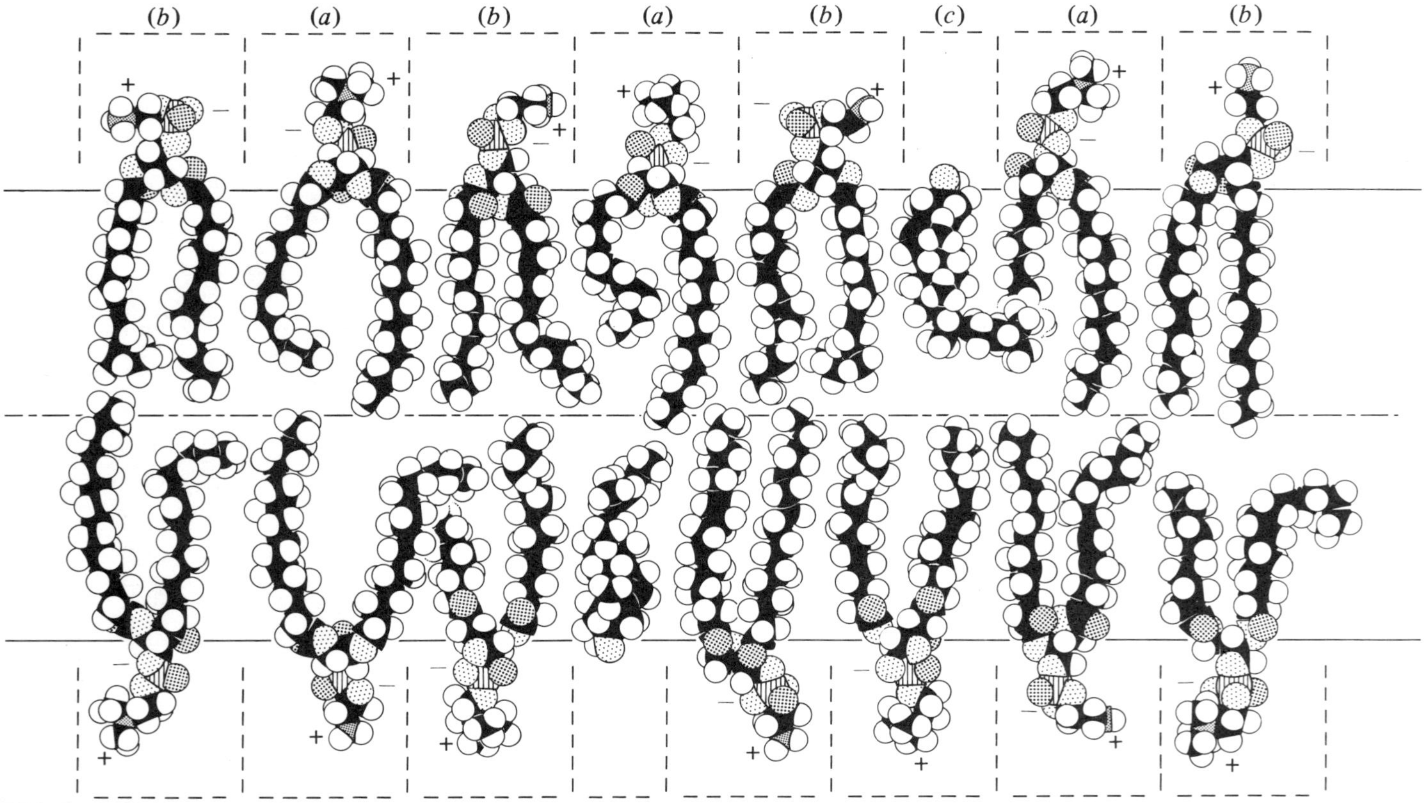

in Fig. 2.18. How these molecules are grouped in relation to each other and to the proteins in the natural membrane is not yet known, though as we shall see in the next chapter, there are good reasons which prevent much cholesterol from packing in the vicinity of phosphatidylethanolamine molecules. The figure also serves as a reminder that the molecules are in constant motion, not only diffusing past each other but also rotating, with the polar groups turning through an angle of 60° and the hydrocarbon chains flexing as they do in a liquid. The effective spaces occupied by the head groups are due to their being kept apart from each other by their charges and their bound water. The polar heads of both phosphatidyl-ethanolamine and phosphatidylcholine can, at times, be almost horizontal in the plane of the bilayer when the positive charge of one is attracted to the negative charge of its neighbour. How these kinds of molecules pack into bilayers is considered in the next chapter.

Suggested reading

Finean, J. B., Coleman, R. & Michell, R. H. (1978). *Membranes and their Cellular Functions*, 2nd edn. Oxford: Blackwell Scientific Publications (Chap. 2).

Quinn, P. J. (1976). *The Molecular Biology of Cell Membranes*. London and Basingstoke: The Macmillan Press Ltd. (pp. 15–75).

3

Structure – cooperative molecules

As we have seen, living membranes, basically bilayers of lipids, contain proteins in different amounts. The proteins range from those superficially and lightly bonded to the surface, through to those buried to varying degrees in the bilayer, some in the hydrophobic region, some traversing the bilayer with water-attracting groups on each side. In this chapter we shall discuss the ways in which lipid molecules pack with each other and with proteins.

The explanations depend on the interaction of several different factors:
  (1) geometry – the size and shape of the different molecules;
  (2) forces – the attraction and repulsion between molecules and particular parts of molecules;
  (3) movements – the movements of different molecules and parts of molecules.

Packing of lipids

How do the lipid molecules of different shapes fit together in a bilayer arrangement, bearing in mind that the lipids on two sides of the bilayer may be different in character and in frequency?

What holds the lipids in the bilayer? The interacting forces of the hydrophilic head groups and of the hydrocarbon chains can be distinguished. Both have attractive and repulsive forces and the result is that, despite their movements, they tend to pack in quite specific ways. Though the space-filling models cannot be regarded as reflecting atomic structure in detail because atoms are not like the little billiard balls, they nevertheless give an accurate idea of the relative spaces which parts of the molecules will occupy and the angles at which the atoms are joined. Space-filling is easily interpreted in relation to the hydrocarbon chains. For instance, a single saturated hydrocarbon chain will, under conditions

42

43

where it is in the solid or crystalline state, straighten out. A single double bond will introduce a perturbation or kink so the chain does not straighten out or pack so effectively (see Fig. 2.4). In practice this means that a lipid with a double bond in its hydrocarbon chain will crystallise at a lower temperature than the corresponding lipid with no double bond. A similar effect is introduced by the branching of a chain. Cholesterol (Fig. 2.5) is particularly bulky and rigid in the rings portion near the —OH group. Consequently, when the hydroxyl group orientates to the water, cholesterol packs rather like a rigid inverted wedge.

Head groups have a profound influence on the different ways in which lipid molecules pack into bilayers; the more important reasons are:

(1) the presence of charges and the possible interaction (attractive or repulsive) of the charged groups;

(2) the hydration shell which will accompany both the charges and any hydroxyl group so that the water molecules, being locally oriented and not in random motion, will fill the space between the covalently bonded atoms of the hydrophilic heads and thus enlarge the effective head-group area (Fig. 3.1);

(3) hydrogen bonds between adjacent hydrophilic heads – particularly important in the glycolipids and, as we shall see, specially significant in galactolipids of thylakoids – may contribute to reduce the effective area of the hydrophilic group.

Fig. 3.1. Water molecules are attracted to the polar heads of lipids, particularly by the charges. Water dipoles align with their oxygens towards the positive charges and their hydrogens towards the negative.

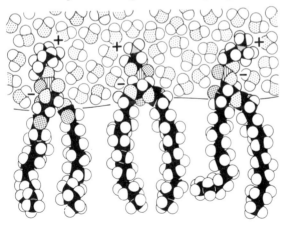

Packing of amphipathic molecules – micelles, bilayers and vesicles

The dynamic shapes of the molecules involved in living membranes govern the packing of the molecules and hence the properties of a local region in the membrane. It is well to understand this in relation to the lipid molecules before approaching the more complex problems of mixed protein-lipid membranes. The principles have been reviewed by Israelachvili, Marčelja & Horn (1980) and related to the thermodynamic principles of aggregation. For our purposes it is sufficient to understand what governs the way in which lipid molecules will pack, as summarised in Table 3.1.

The molecules we are discussing are held in the bilayer of biological membranes by their strong tendency, at the head-group end, to be pulled to the water on each side, counteracted by the hydrophobic chains' strong tendency to escape from the water into each others' presence. If the effective head-group cross sectional area (taking account of charge, hydration etc.) is large compared with that of the hydrophobic chain cross section area, we would expect a different packing of the molecules from the situation in which the head-group area is small and the hydrocarbon chain area is large. Indeed, the different shapes which surface-active substances assume in both natural and artificial membranes is largely due to this property. There is an optimum surface area per molecule defined at the hydrocarbon water interface at which the total interaction energy per molecule is at a minimum. The other important characteristic of these molecules is that, though they may be fluid, they are incompressible and the chain cannot be extended beyond a maximum length. These factors govern the packing (Fig. 3.2).

Wide head groups, narrow hydrocarbon chains: micelles

If the radius of the optimal area is sufficiently large and the hydrocarbon volume is sufficiently small, the molecules with heads in water and tails towards each other, will form spheres – known as *micelles* (Table 3.1a, Fig. 3.3). The average packing shape of the molecules is that of a wedge. Most molecules which form micelles have large head-group areas because they are charged, and have only one hydrocarbon chain. The number of molecules which can pack into the micelle will be varied if the effective head-group area is altered. An interesting example is provided by sodium dodecylsulphate. The effective head-group area is due to the electrostatic head-group repulsion. If the molecules of sodium dodecyl-sulphate are transferred from water (where the head groups have maximum

Table 3.1. *The packing characteristics of the different lipids and the structures they form. From Israelachvili et al.* (1980)

Packing shape	Lipid types	Structures formed
a Cone or wedge	Single-chain lipids head area: large (some lysophospholipids NaDS in low salt)	Spherical micelles
b Truncated cone or wedge	Single-chain lipids head area: small (non-ionic lipids, lyso-lecithin NaDS in high salt)	Globular or cylindrical micelles
c Truncated cone	Double-chain lipids head area: large fluid chains (lecithin, sphingomyelin, phosphatidylserine in water, phosphatidyl-glycerol, phosphatidic acid, disugardiglycerides, some single chained lipids with very small, uncharged head groups)	Flexible bilayers vesicles
d Cylinder	Double-chain lipids head area: small (anion lipids in high salt, saturated frozen chains, phosphatidylethanolamine, phosphatidylserine$+Ca^{2+}$)	Planar bilayers
e Inverted truncated cone	Double-chain lipids head area: small (non-ionic lipids, poly (cis) unsaturated chains, highly unsaturated phosphatidyl-ethanolamine, cardiolipin $+Ca^{2+}$, phosphatidic acid$+Ca^{2+}$, monosugar diglycerides, cholesterol)	Inverted micelles

Fig. 3.2. Like molecules can aggregate and crystallise (a), e.g. as temperature drops and fluidity decreases. Such molecules (a) lengthen and narrow compared with the adjacent fluid molecules (b), so differences in thickness of the membrane result.

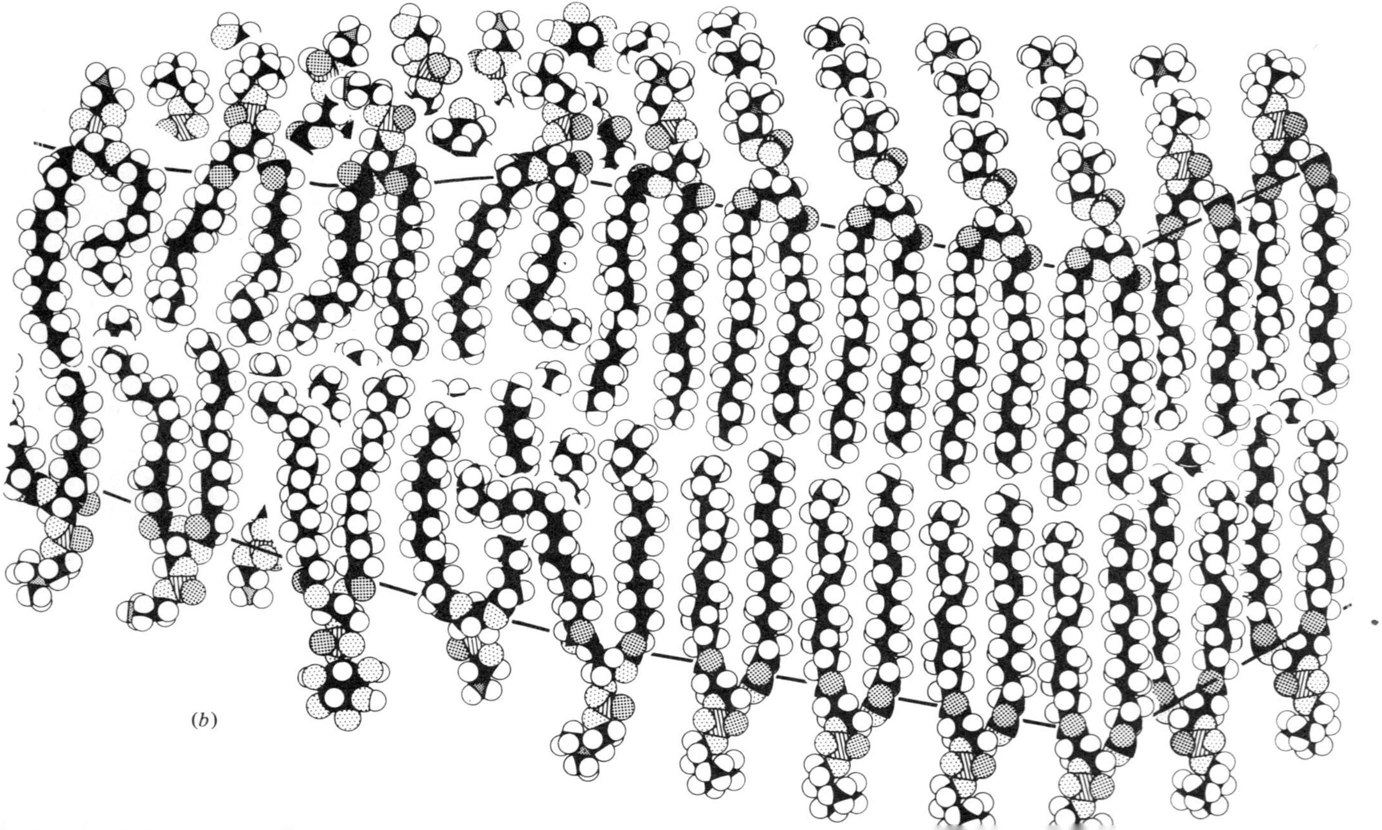

(b)

47

electrostatic mutual repulsion) to a high concentration salt solution, the
ions of the salt screen the repulsive charges between the micelle molecules
and the effective head area is reduced. This means that, with the smaller
head area, more hydrocarbon chains (themselves of constant small area)

Fig. 3.3. Molecules packing into a micelle, shown in cross section. The
effective head group has larger area than the single hydrocarbon tail
because of the charges repelling each other. The hydrocarbon tails of
the adjacent molecules are also shown.

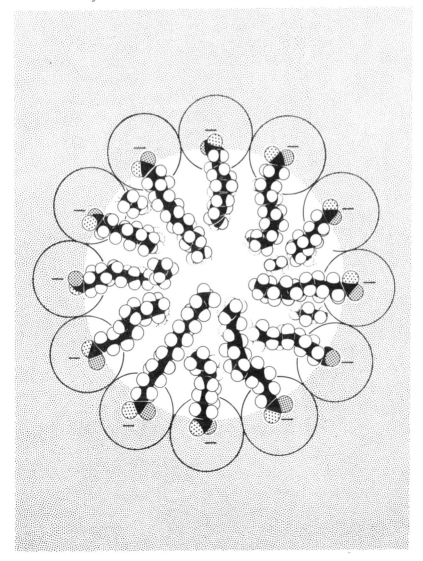

can pack. Thus, small micelles of sodium dodecylsulphate in water with only about 60 molecules become large micelles with about 1000 molecules in high concentration salt. In addition to sodium dodecylsulphate, some naturally occurring lysophospholipids (which have only one hydrocarbon chain each) can form micelles, a property which results in sharply curved regions in the bilayer which, as we shall see, may have biological importance.

Some lipids which have small head area because they are non-ionic, have single chains. The area of the head still exceeds that of the hydrocarbon chains but the packing shape now becomes that of a truncated cone or wedge. These molecules do not pack as tightly as those which form micelles, as we have seen with sodium dodecylsulphate in a high concentration of salt. Such non-ionic lipids and also lysolecithin form larger micelles which are globular or cylindrical (Table 3.1*b*).

Wide head groups, double hydrocarbon chains: flexible bilayers and vesicles

If the head groups have a large area, the double chains of the lipids confer a wide area on the hydrophobic part of the molecule, and a more nearly planar structure will be formed and the molecules will pack into a bilayer (Table 3.1*c*). The packing shape becomes that of a truncated cone approaching a cylinder and, if the hydrocarbon chains are fluid, i.e. do not crystallise to form a solid structure, a flexible and often curved bilayer is formed. Thus many of the naturally occurring lipids form vesicles or *liposomes* – a property which has been greatly used experimentally to increase our knowledge of the properties, not only of membranes but also of membrane-bound enzymes (see Bangham, 1980). In this category are the vesicle formers listed in Table 3.1*c*. Sometimes, when the polar head is particularly wide, as in sphingomyelin and some digalactolipids, one hydrocarbon chain needs to be longer than the other and, to some degree, coiled to fill the packing space which would otherwise be left under the head group. This is illustrated by the coiled tail of a cerebroside, as seen in Fig. 2.9.

Narrow head groups, double hydrocarbon chains: planar bilayers

The logical extension of the solid geometry of these molecules shows that double chain lipids with small head-group area would be cylindrical in shape. Such molecules, when packing together, would not have any tendency to form a curved structure. Consequently a planar bilayer is formed and is the kind to be expected of the phospholipid, phosphatidylethanolamine (Table 3.1*d*).

49

When a divalent cation (e.g. Ca^{2+}) binds two molecules of the emulsifying agent (A), the area of the hydrocarbon chains is greater than the hydrophilic head group and the curvature is reversed so the water (C) is trapped in droplets in the oil (B).

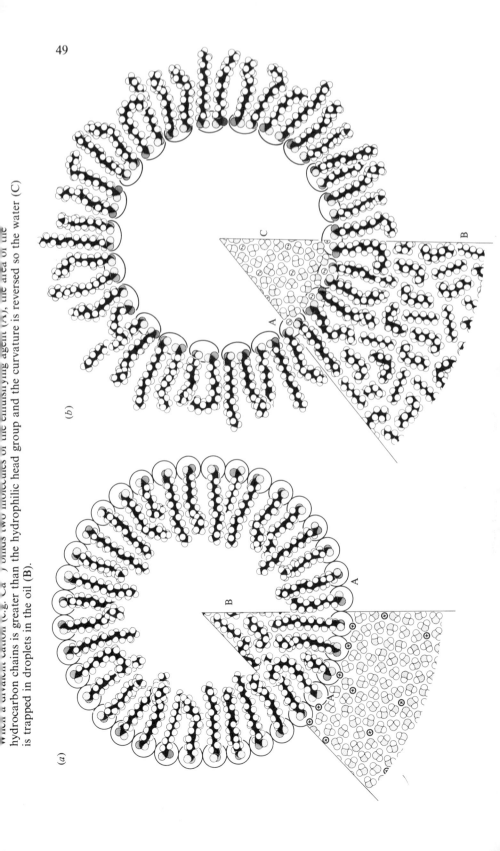

(a)

(b)

Small head-group areas, wide hydrocarbon chains: inverted micelles

Inevitably, if the head-group areas are very small and the hydrocarbon-chain areas are large, the packing shape will become that of an inverted truncated cone (Table 3.1e). This would mean that the formation of micelles becomes inverted, i.e. the water attracted to the hydrophilic heads will be entrapped in small spheres and the lipophilic tails will satisfy each other. An interesting analogy is to be seen in the simple experiment of an oil-in-water emulsion where the emulsifying agent is a sodium salt of a soap. The monolayer between the oil in the water is curved because of the area of head group and lipophilic tails of the soap, and oil droplets are enclosed to give an 'oil in water' emulsion (Fig. 3.4a). If the sodium is replaced by Ca^{2+}, the negative charges of two soap molecules are attracted to the divalent ion and the two hydrocarbon chains, now wider than the head group, impose a reverse curvature which entraps water droplets, and the emulsion becomes 'water in oil' (Fig. 3.4b).

So far we have been talking about the packing of lipid molecules into different arrangements as though it were a static phenomenon but we must remember that these molecules are not stationary and are moving in various ways. This will be dealt with more fully in the next chapter but some aspects, relevant to the packing, should be mentioned here.

Self-assembly of lipid structures

As we have seen, the packing shapes of the molecules will govern the way in which they pack into different structures, but what determines that they will pack at all? This is related to energy; molecules will come together if the free energy per molecule, in the aggregated state, is less than that of the dispersed state. The kind of structure formed in aggregation, whether micelle, bilayer or precipitate, is also dependent on the free energy per molecule; particular molecules aggregate into the structure in which the free energy per molecule is less than that of any other structure. Put in another way, it is less likely that lipid molecules will remain in water where they are relatively insoluble if they can group together with the hydrophobic forces of the hydrocarbon chains satisfying each other and the water molecules satisfying their own electrostatic attraction and hydrogen bonding with the hydrophilic groups. As lipid molecules are not completely insoluble in water, there will always be a balance between the concentration of molecules in the water and those in the aggregate or micelle. This also applies to the separation of an oil from water. Oil molecules would be present in water as monomers with a few dimers, trimers and so on, but as the association of oil to form a bulk phase, separate from the water, is so strongly favoured, aggregation just goes on until all but a few molecules of the oil, which remain dissolved in the water, are associated with each other to form the bulk oil. So it is with aggregate formation, and it is useful to speak of the 'critical micelle (aggregate) concentration' which corresponds to the solubility of the individual lipid molecules in the aqueous phase. The critical concentration for different lipids is revealing in telling us about the properties of the system. The critical concentration for micelle formers (single chains of 12–16 carbon atoms) is about 10^{-2} to 10^{-3} M. By contrast, the critical concentration for bilayer-formers with double-chains of 14–18 carbon atoms is only about 10^{-10} M, i.e. about 10–100 million times less concentrated. We shall return to discuss the average times that individual molecules stay in the membrane (the residence time) when we are discussing molecular movements from solution to aggregates.

We may think of the packing more precisely when we remember that the relative dimensions of the molecules concerned govern the shape and size of the aggregates. A useful contrast comes from a comparison of egg lecithin, which forms vesicles, and lysolecithin, which forms micelles. Israelachvili *et al.* have calculated the importance of the critical 'packing parameter' $V/a_o l_c$ where V is the hydrocarbon volume of the molecule, a_o is the area of the head group and l_c is the length of the hydrocarbon chain. The larger the critical packing parameter, the larger the aggregate body. Using their equations and assuming that the head group area of egg lecithin is about 0.717 nm², the volume is 1.063 nm³ and the length is 1.75 nm, the critical packing parameter is about 0.85. The radius of the vesicle would be about 10.8 nm and the total number of molecules about 3000. Lysolecithin would have about the same head-group area and the same chain length but only half the volume, so the critical packing parameter would be about 0.42. Such a packing should result in the formation of small globular (non-spherical) micelles containing about 186 molecules.

The packing parameter will be varied by other factors. For example, introducing smaller head-group areas will lead to larger vesicles or less curved bilayers. The introduction of increased unsaturation, particularly of *cis* double bonds into the chains, decreases chain length which, by increasing $V/a_o l_c$, would also cause larger vesicles. Increase in temperature, which would increase the mobility of the hydrocarbon chain, will reduce the effective length, l_c and, by increasing $V/a_o l_c$ would also lead to larger vesicles. Similar factors are obviously involved in determining the thickness of the bilayer.

Mixed lipid bilayers

So far we have considered the packing properties of individual lipid molecules but, as we have seen, biological membranes contain a mixture of lipids and therefore a mixture of different shapes. Thus the different lipids will compensate for each others' differences, e.g. a truncated cone could pack next to an inverted truncated cone.

Sometimes, some molecules in a portion of a membrane may separate out into a definite phase because their mutual attraction is strong enough to overcome the kinetic movement keeping them in the dispersed state. Sometimes a similar separation can occur if the attractive forces between the polar head groups are increased. For example, Ca^{2+} can pull together anionic lipid molecules and cause them to separate out (Fig. 3.5).

The mixed packing will depend on the principles already discussed. If lysolecithin were added to lecithin vesicles, the result would be smaller vesicles with more highly curved bilayers. If enough lysolecithin molecules are added to the mixed bilayer to associate with each other, their greater curvature can so distort a local part of the bilayer that a pore may be formed (Fig. 3.6). A spectacular increase in permeability to ions and small molecules would result. Phosphatidylethanolamine and cholesterol, which both have the shape of inverted cones (Table 3.1e), if mixed with lecithin would increase vesicle size and decrease the curvature, or even make it go the other way. Only limited amounts could be added before the bilayer and the vesicle structure would be destroyed. Cholesterol and lysolecithin, in certain proportions, can associate (their shapes being complementary) to form a bilayer. However, not surprisingly, natural membranes never contain high amounts of both cholesterol and phosphatidylethanolamine

Fig. 3.5. Effect of a divalent ion on adjacent lipid molecules. The calcium ion (Ca^{2+}) attracts and holds two negative charges on adjacent phospholipids; fluidity in the hydrocarbon chains decreases, they increase in length as van der Waals attraction between them overcomes the kinetic movement and the membrane thickens.

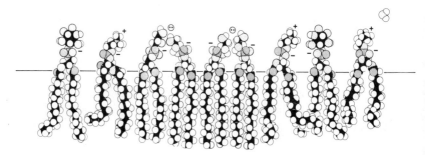

53

Fig. 3.6. Consequence of an effectively wide head group with a narrow hydrocarbon chain in a membrane. Curvature must increase and pores (*a*) may be formed.

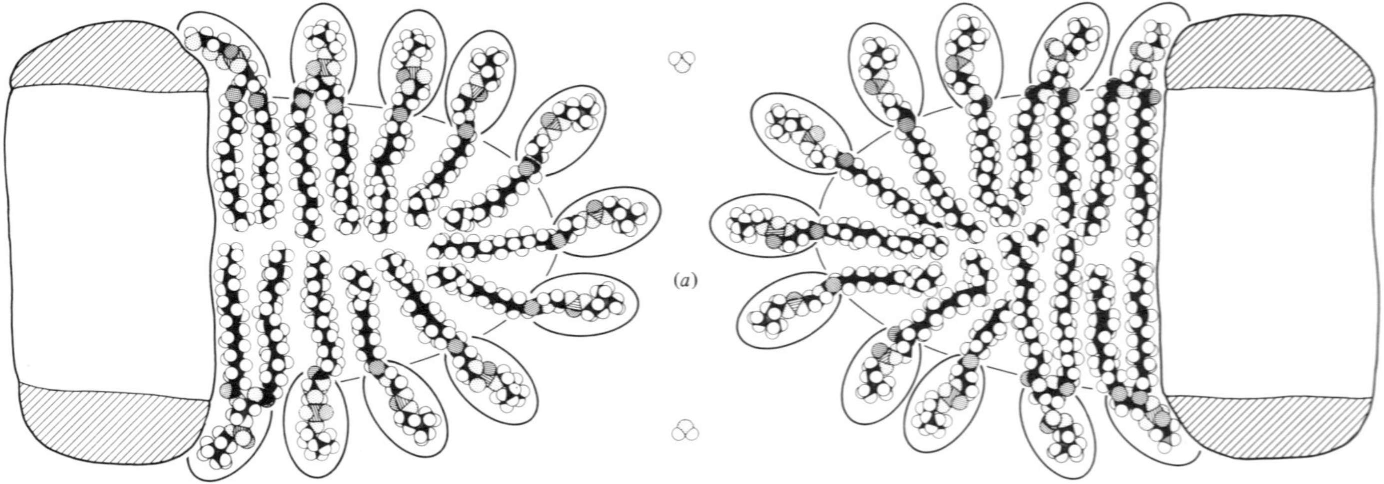

together in one layer. The membranes of erythrocytes contain both, but they have the two lipids on opposite sides of the bilayer. The rigidity of cholesterol molecules is best compensated by the fluidity of the lipids and, at high concentrations, cholesterol molecules are separated from each other by at least one fluid lipid molecule. A rigid molecule like cholesterol may result in a straightening of the adjoining hydrocarbon chains, which is accompanied by a thickening of the bilayer and *reduced* fluidity – a lesson for considering the effect of protein packing in mixed bilayers (Fig. 3.7).

Some small lipid-soluble molecules have the opposite effect to cholesterol. By dissolving in the hydrocarbon region of the bilayer they *increase* the fluidity and decrease the rigidity of the structure. If enough dissolved molecules penetrate to the centre of the bilayer they may increase the thickness of the membrane. They are most likely to interfere with the attachment of lipids to proteins and of lipids to lipids. Their presence may move the lipid molecules further apart and result in proteins being pushed further away from each other (Fig. 3.8). This separation of protein molecules may inhibit a process, e.g. a multiple enzyme reaction, which depends on a contact between adjacent protein molecules. Many people believe that general anaesthetics – those small lipid-soluble molecules – act on nerve processes by upsetting the membrane organisation in this way. We shall return to this interesting and important effect later.

Fig. 3.7. The rigid cholesterol molecule (*a*) can immobilise the adjacent phospholipids' hydrocarbon chains as the van der Waals attraction overcomes their kinetic movement. The molecules lengthen, narrow and, as a result, the membrane thickens.

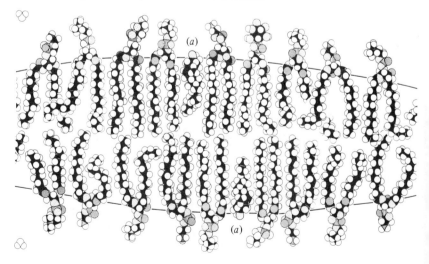

55

Fig. 3.8. A lipophilic small molecule (e.g. ethanol) may dissolve in the centre of the bilayer and produce several effects: (a) the membrane before the small molecule enters; (b) the membrane after: fluidity is increased, molecules, including proteins, may be moved apart, lipids may be detached from their normal position on protein surfaces and, despite increased fluidity, the membrane may thicken due to the volume of the small molecules trapped.

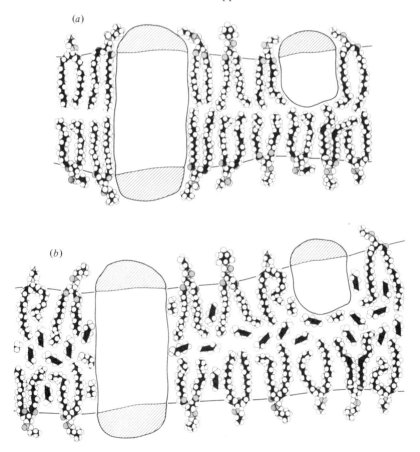

(a)

(b)

56

Proteins in mixed bilayers

As we have seen, most proteins are structurally rigid. When they are surrounded by lipids, those near the proteins are unable to move as freely as they did in the uninterrupted presence of other lipids. The hydrophobic forces are, therefore, strong enough at protein–lipid interface to make the proteins appear to be bound to adjoining lipids. Furthermore, the structural rigidity in these lipids will affect those in the immediate vicinity. The lipids adjoining the proteins have variously been termed 'boundary lipid', 'annulus' and 'halo' (Fig. 3.9). Such perturbed lipids in the vicinity of the proteins occur with cytochrome c when bound to the membrane, with cytochrome oxidase, with rhodopsin and bacterio-rhodopsin and with various other membrane proteins.

As might be expected, the association of some lipids with some proteins is fairly specific; with other proteins the association is much less specific, with weaker forces holding the lipids to the protein. We have discussed the

Fig. 3.9. The ring of lipid molecules tightly attached to a protein. The bonds may be polar or hydrogen in the head groups, are van der Waals in the hydrocarbon area, with the possibility of an occasional sulphydryl linkage. The hydrocarbon chains will conform to the shape of the adjacent protein surface.

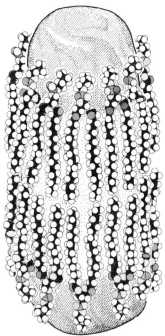

protein, cytochrome oxidase; it normally has six to ten associated lipids. They do not exchange with the other lipids of the bilayer and, as they co-purify with the protein, can be regarded as bound. Since negatively charged lipids are preferentially bound to the protein except in high salt, which decreases the electrostatic interaction and abolishes the preference for anionic lipids, both the electrostatic forces and the hydrophobic forces of protein and lipid are involved. Though proteins are rigid, they have irregular surface shapes and the fluid lipid chains must deform to fit with the protein hydrophobic surfaces within the limits of maintaining lipid surface area at near optimum value and of not extending beyond their maximum length (Fig. 3.10). Though the lipids in the vicinity (and even those on the other side of a bilayer) may be affected by the distortion, the effect decreases at two to three lipid diameters from the surface. This kind of packing is best done with fluid lipid chains; indeed, the less fluid the lipids, the more difficult it is to get proteins to pack into the bilayer. That rigid molecule, cholesterol, for instance, tends to inhibit the packing of proteins. Thus it is not surprising that most of the protein of erythrocytes is on the inner, fluid side of the bilayer, not on the outer cholesterol side.

Peripheral proteins such as cytochrome c can interact with the membranes by electrostatic attractions, both with the proteins and with the lipids in the environment. Thus the positive lysine groups of cytochrome c, combining with the proteins in the surface, are likely to affect the nearby lipids which also have negative charges.

We have seen that the protein content of living membranes is quite high, e.g. 74% in the mitochondrial membrane. This means that there are

Fig. 3.10. The principle of interaction between a hydrocarbon chain and portion of an α-helix: (a) the lipophilic side chains of alanine and phenylalanine on the lipophilic protein; (b) the hydrocarbon chain; (c) the two held together by hydrophobic forces.

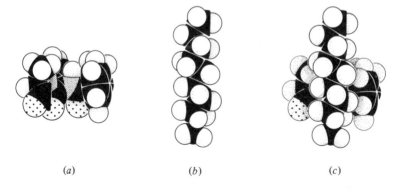

(a)   (b)   (c)

probably few lipid molecules in living membranes which are not directly or indirectly under the influence of proteins.

In subsequent chapters we shall consider the molecular movement to be expected of the different substances in membranes and the importance of their packing and movement on the function of membrane systems.

Suggested reading

Bangham, A. D. (1980). Development of the liposome concept. In *Liposomes in Biological Systems*, ed. G. Gregoriadis & A. C. Allison, pp. 1–24. New York: John Wiley & Sons Ltd.

Israelachvili, J. N., Marčelja, S. & Horn, R. G. (1980). Physical principles of membrane organisation. *Quarterly Review of Biophysics*, **13**, 121–200.

Lodish, H. F. & Rothman, J. E. (1979). The assembly of cell membranes. *Scientific American*, **240**, 38–53.

Quinn, P. J. (1976). *The Molecular Biology of Cell Membranes*. London and Basingstoke: The Macmillan Press Ltd. (pp. 80–112).

4

Dynamics – moving molecules

I have stressed, several times, the importance of the movements of molecules and, in this chapter, we shall see that those movements which are critical to the function of membranes, are also different in different circumstances.

The bilayer molecules are confined to the membrane and, except for some of the peripheral proteins, tend to remain in the membrane. However, as all molecules are moving, there is a range of probabilities for individual molecules to escape from the membrane and for individual molecules in the surrounding water to rejoin. This happens with low frequency and, on average, not sufficiently often to disrupt the membrane structure. Membrane molecules perform characteristic movements in the plane of the membrane.

Evidence for movement of membrane molecules

We know how molecules move in membranes because of the adaptation of several modern physical techniques, notably NMR, ESR, X-ray diffraction and scanning calorimetry. With the NMR technique, a measurement is made over about 10^{-5} s, during which time a molecule may assume a large number of conformations and might diffuse through a large distance, e.g. 50 nm. However, despite its slowness, the technique is valuable when hydrogens of the chains are replaced by deuterium at specific places. Results can be interpreted to show that the movement along a chain remains relatively constant, increasing near the end in the middle of the bilayer. The ESR technique has a much greater sensitivity and movements as fast as about 10^{-10} s can be deduced. It suffers from the disadvantage that the ESR probe or spin label itself is likely to introduce a perturbation into the membrane.

Experiments with X-ray diffraction show that, at higher temperatures,

59

the hydrocarbon chains of the lipids give broad diffraction bands similar to those of liquid paraffins, indicating that they are in a liquid state. At lower temperatures, motion is lost and the X-ray reflection corresponds to stiff, fully extended chains with only rotational motion.

Complementary information is obtained from scanning calorimetry which can measure the transition temperature, i.e. the temperature at which the hydrocarbon chains pass from the liquid to the solid state and thus provide information about the interaction forces between the phospholipid molecules and their interaction with the aqueous phase. Our knowledge of these movements has been built up by using the information from the different techniques, each of which has some disadvantage, and combining it with appropriate theoretical treatment.

Diffusion (in water) versus aggregation: flip-flop

We have discussed the way in which a critical micelle concentration of a lipid molecule in water will be related to the nature of the aggregate and how micelle-formers have a much higher water solubility than bilayer-formers. The lower solubility of bilayer-formers in water when in contact with the large, many molecule aggregates they form, is due to their higher hydrophobic energy. As the molecules in the micelle and bilayer are subject to the diffusive forces of kinetic movement, they will be coming and going from the water phase. As might be expected, the higher hydrophobic energy associated with bilayer-formers gives their molecules more chance of remaining in the bilayer for longer periods than those of micelle-formers. Various authors, see for example Israelachvili *et al.* (1980), use the concept of 'residence time' for such molecules and it is found that, for micelles, it is about 10^{-4} s. By contrast, that for a molecule in the bilayer is 10^4 s, i.e. about 3 h or 100 million times longer than in micelles.

The same forces which make for long residence times also work against *flip-flop*, i.e. the chance of a molecule in one side of a bilayer flipping over to the other side of the same bilayer. Flip-flop times work out at 10^4 to 10^5 s for lipids, i.e. of the order of hours or days. However, although this kind of movement is so restricted in pure lipid bilayers and would be similarly restricted in any pure lipid domain in the living membrane, it is completely changed in the vicinity of some of the other membrane molecules and especially near proteins, where the rate of flip-flop can be increased by several orders of magnitude.

Significance of molecular movements in membranes

Many people have difficulty in appreciating the dynamic nature of molecules in membranes. This difficulty arises for several reasons. Though everyone has, for generations, accepted the idea that the molecules of both solute and solvent are in constant movement with collisions which are the basis of reactions, dimer formation and so on, the two dimension bilayers, particularly as first presented, introduced a static picture. In the years after the Davson and Danielli model was described, the static diagram of the bilayer like two picket fences, foot to foot, suggested a rigidly organised structure and we rarely made clear the molecular movements which must be taking place. Many people used to question me as to how such a structure with interactions between those molecules (or tacks between the pickets) could allow diffusing molecules to penetrate and, especially, a favourite doubt, how could proteins, e.g. those of secretory organs, pass across such a membrane? I used to reply that we had only to visualise a tank coming to a garden fence which would be flattened by its passage. Now, of course, we know that a massive set of forces such as a protein molecule may introduce such a perturbation at the membrane surface that the temporary localised disruption could allow the protein through. It is, therefore, necessary to spend some time getting into our minds the very dynamic picture of membrane molecular movements – for some molecules, as dynamic as those in solution but confined to the bilayer. Such ideas are difficult to convey by static diagrams.

Within the confines of the bilayer we therefore recognise the usual movements of molecules:

(1) rotation of molecules within the plane, but not much tumbling or flipping at right angles to the plane;

(2) diffusion within the bilayer. The lateral movement of molecules is rapid, similar to molecular diffusion in solution but confined to the two dimensions. It has been calculated that a molecule of lipid can diffuse the length of a bacterial cell in 1 s. Smaller molecules which are not lipids or proteins but which have some solubility in the membrane will also diffuse. By contrast, the boundary lipids adjoining a membrane protein will exchange with the other lipids only more slowly;

(3) fluid movement of the hydrocarbon chains. As we have seen, the hydrocarbon chains at ordinary temperature are usually quite fluid due to the rotation about single bond —C—C— linkages. This property has many effects, not only on the packing of the

molecules into the bilayer and on its thickness but also on the movement of other molecules in the hydrophobic region;

(4) movements at right angles to the plane of the membrane, or bobbing up and down without flipping. This has profound importance in relation to movement of ions into and across the bilayer, in proton translocation of light-trapping and oxidation–reduction membranes and probably in some enzymatic reactions. It can apply to both protein and lipid molecules;

(5) conformational changes in proteins. Substantial movements of parts of proteins can occur as the results of the chemical changes associated with enzyme–substrate complex formation and, in membrane-bound enzymes, these may be important as one of the controls of enzyme action.

Molecules rotating

The rotational movement of molecules occurs everywhere in fluid situations and is slowed down markedly as the molecules pass into a crystalline or solid state. Nevertheless for most of the time, we can expect rotation of most of the molecules including the proteins in a living membrane; but because of the constraints of the water-attracting groups, the rotation is in two dimensions in the plane of the membrane.

Rotation in the other plane, i.e. flip-flop is much less likely. However, as we have also seen, these molecules are not simple homogeneous cylinders and the type of rotation will vary in different parts of the molecule, especially due to the fluid nature of the hydrocarbon chains. The rotation of molecules in membranes offers one of the control mechanisms that the cell may use in multienzyme reactions. For example, a protein molecule may have an enzyme complex reaction on one side and then rotate in its changed form to react with another enzyme on its other side. It is worth noting that the hypothetical molecule or 'carrier' or 'protein' or 'permease' which, it was realised must be membrane-bound and specifically transporting a molecule or an ion across, might pick up the molecule on one side and rotate in the membrane to liberate it on the other. Though this is not impossible because of perturbing effects of proteins, it is unlikely that a properly membrane-bound protein would be able to do so because of the constraints against flipping. As we shall see, there are other mechanisms which must be looked at for permease activities.

Lateral diffusion in the bilayer

As we begin to get the picture of a natural bilayer as an ever-changing, seething mass of lipids, proteins and other lipophilic molecules in which normal molecular movements are occurring in two dimensions, it is easier to appreciate the significance of living membranes (Fig. 4.1). All molecules will be capable of diffusing past each other in the membrane unless, as can happen with fall in temperature, the intermolecular attractive forces become greater than the kinetic translational forces. If the temperature drops the characteristic process of crystallisation can occur and like molecules will come together in solid domains within the membrane while unlike molecules are excluded (see Fig. 3.2). Thus the lipids undergo a liquid-crystalline to gel phase transition with decreasing temperature, an important property which will be discussed elsewhere. However, in a mixed bilayer, so long as the fluid tendencies exceed the crystalline tendencies, the molecules will diffuse. It is likely, given the structural heterogeneity within membranes, that different molecules can diffuse at very different rates, even though they may be of similar size. Thus a molecule may diffuse between two membrane-bound enzymes and react successively with both. As we shall see, the two essential membrane-bound molecules, plastoquinone in photosynthetic membranes and ubiquinone in respiratory membranes can behave just this way. The diffusion of plasto-quinone is represented in Fig. 4.2.

Small molecules which dissolve in some parts of the bilayer will also diffuse laterally, the most likely small molecule being that which is soluble in the lipophilic region of the fluid hydrocarbon chains. There it can diffuse laterally within the hydrophobic region or diffuse to the other side of the bilayer. Its ingress and egress at the bilayer depend, to varying degrees, on its acceptability to the head group region. We shall return to the diffusion of such molecules in Chapter 7.

Water and non-electrolytes, in consequence, probably cross the interior of a bilayer in the same way as they diffuse through an oil in which they may be only sparingly soluble. The solubility of water in various hydrocarbons is of the order of 10^{-4} M. There are, therefore, always some water molecules (about 1 per 20–40 lipid molecules) in the hydrocarbon and an alteration of concentration of water (e.g. osmotic changes) on one side of the bilayer results in rapid diffusion to equilibrium on the two sides (Fig. 4.3).

Ions do not tend to enter the liquid hydrocarbons to any extent; in partitioning, almost all the ions remain in the aqueous phase. For a typical unhydrated ion of radius 0.1 nm, the partition coefficient is about 10^{-60}.

64

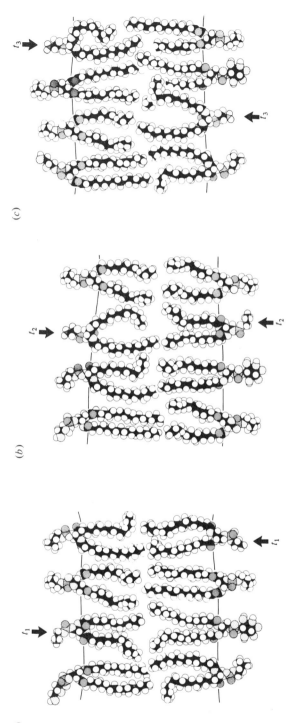

Fig. 4.1. Lipid molecules are in constant diffusion in the plane of the bilayer. The phosphatidylethanolamine in the upper layer, marked with the arrow, is shown as moving from left to right past other molecules at three intervals of time, $t_1(a)$ $t_2(b)$, and $t_3(c)$. The phosphatidylethanolamine in the lower layer is shown as moving from right to left.

65

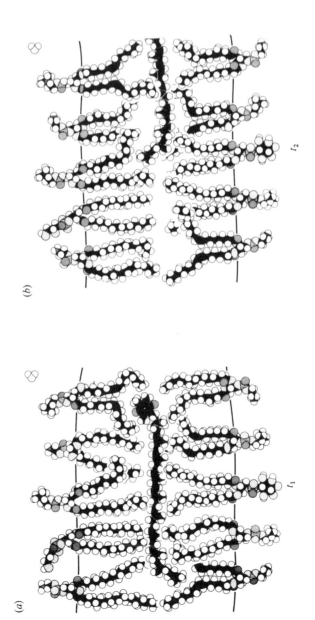

Fig. 4.2. Plastoquinone, which, especially in its uncharged forms (quinone or hydroquinone), is very soluble in the fluid part of the bilayer, diffuses readily between the lipid tails in the time interval t_1 to t_2: (a) t_1, (b) t_2 with the molecule diffusing to the right.

However, the hydrocarbon interior of hydrated phospholipid bilayers is more polar than a pure hydrocarbon. Whereas a pure hydrocarbon phase has a dielectric constant of about 2, experiments show that the dielectric constant in a bilayer is above 20 down to the C_5 position and about 5.5 at the C_{12} position. The effects of this on ions entering the bilayer (assuming they have passed an unfavourable field of charge at the head groups) is complicated. An ion can enter such a region, surrounded by a shell of water molecules, as a hydrated ion, and it can be calculated that this will make a very big difference to the number of ions that can be in the hydrocarbon phase (Table 4.1). The permeability of ions which form hydrated ions with

Fig. 4.3. Individual water molecules or small groups of two to three molecules dissolve in hydrocarbon and move rapidly in a fluid layer. An osmotic difference on the two sides of the membrane due to sugar (lower side) results in water movement from one side of the membrane (top) to the other (bottom).

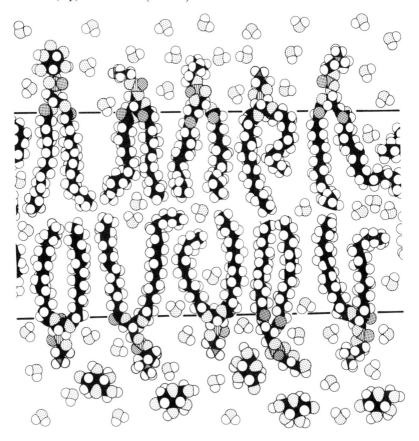

water molecules is independent of the charge of the ion, which probably explains why different ions traversing lipid membranes have similar permeabilities. We shall return later to discuss the range of ionophores formed by substances, other than water, which have various degrees of specificity for ions and which take ions across the membrane; some of these, like valinomycin and nonactin, owe their permeability in membranes to the lipophilic properties of the outside of the ionophore molecule.

Table 4.1. *Ions and hydrated ions in pure hydrocarbon and bilayer hydrocarbon. From Israelachvili* et al. (1980)

Assumed dielectric constant for hydrocarbon	Example	Concentration in hydrocarbon	
		Ions	Hydrated ions
2	Pure hydrocarbon	$10^{-61} \times$ concn in water $r = 0.1$ nm	$10^{-26} \times$ concn in water $r = 0.36$ nm
6	Bilayer interior	$10^{-22} \times$ concn in water $r = 0.1$ nm	$10^{-12} \times$ concn in water $r = 0.25$ nm

r = radius of ion or hydrated ion

Fluid movement of hydrocarbon chains

This movement is of great importance for several reasons:

(1) it allows for the mobility of adjacent molecules, e.g. proteins;

(2) it allows hydrocarbon lipids to accommodate better to the irregular hydrophobic surfaces of proteins;

(3) it gives a fluid region in the membrane for the lateral movement;

(4) it allows the lipid molecules to occupy a larger cross sectional area, increases their packing area and reduces the total width of the bilayer. By contrast, if the movement ceases, e.g. with lowered temperature, the molecules' volume remains the same but the cross sectional area decreases, the molecules pack tighter and the bilayer becomes thicker (Fig. 4.4);

(5) it allows for a differential in the tightness with which hydrocarbon chains pack along the length of the chain. This is of great importance for it allows for a bilayer to have a fluid structure in the centre of the membrane with semi-solid or quasi-crystalline structure in the same chains near the hydrophilic heads (Fig. 4.5).

This important property requires some explanation. If the attractive forces of the head groups are strong, the movement of the hydrocarbon chains will be restricted in the immediate vicinity, especially as there is almost always no double bond in the chain until carbon atom 8. The attractive forces of the head groups hold the molecules sufficiently for the attractive forces between the adjacent $—CH_2—$ chains to overcome their kinetic movement. However, further along the molecule, away from the polar group and especially if there is a double bond at carbon atom 8 introducing a 'kink', the fluid properties of the molecule are preserved. Thus it is possible for the chains near the head groups to be in a quasi-solid state when the centre of the membrane has the properties of an isotropic liquid. The functional importance is considerable: it enables a membrane to 'trap' molecules which can have considerable lateral diffusive movement in the centre of the membrane but escape only with difficulty through the tight hydrocarbon chains which reduce the permeability.

Fig. 4.4. A bilayer of phosphatidylcholine and phosphatidylethanolamine as temperature is lowered from T_1 (a) to T_2
(b) to T_3 (c) and the membrane tends to solidify. The molecular fluidity
decreases; like molecules aggregate with like molecules in each layer,
the hydrocarbon chains lengthen and the membrane thickness
increases. The solidification temperature is quite sharp – like a melting
point.

(a)

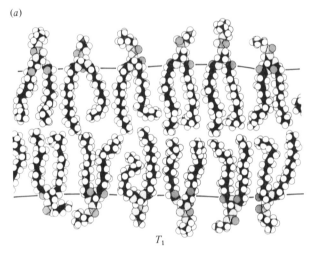

T_1

(b)

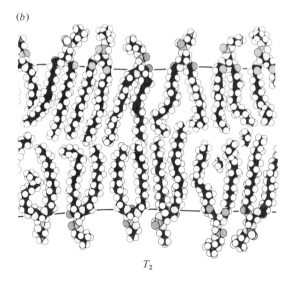

T_2

70

Fig. 4.4. (*cont.*)

(*c*)

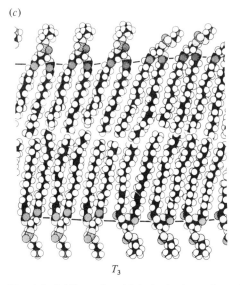

T_3

Fig. 4.5. A bilayer in which the hydrocarbon chains are in the solid state (A) near the polar groups and, due to double bonds at the eighth carbon atoms, are in the fluid state (B) in the centre of the bilayer. Small lipid-soluble molecules (C) in the centre can diffuse rapidly in the plane of the membrane but are prevented from escaping to either side.

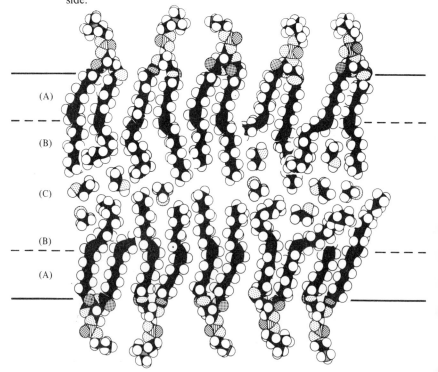

Vertical movements: bobbing up and down

Though any membrane molecule can leave only with difficulty, the depth at which it is found in the bilayer is not constant. For example, if a change in the polar head of the molecule results in extra charge (as it might do by addition of a proton or electron), its hydrophilic attraction will be increased and it will move further into the aqueous phase between the other head groups. We believe that this is the way in which the molecules of reduced plastoquinone and ubiquinone move partly out of the bilayer, 'bobbing up' into the aqueous phase (Fig. 4.6). When one of these molecules picks up a proton (if it has a single negative charge) or two protons (if it has two negatives), it immediately loses some of its water attraction. The very strongly hydrophobic tails dissolved and moving actively in the central hydrophobic zone, pull the molecule back in or cause it to 'bob down'.

The lipid molecules themselves would also be affected by any change in their hydrophilic attraction. Phosphatidylcholine appears to combine with a chloride ion (at the positively charged choline) and a proton (at the negatively charged phosphate). When this happens the charge is decreased, the water attracting tendency decreases and the phosphatidylcholine sinks into the bilayer (Fig. 4.7). This was first postulated because Pagano & Thompson (1968) noticed that if $^{36}Cl^-$ is put into a KCl solution on one side of a phosphatidylcholine bilayer, the Cl^- and $^{36}Cl^-$ exchange across the bilayer about 1000 times faster than KCl can diffuse across. Toyoshima & Thompson (1975a, b) suggested that the combination of Cl^- with the choline and of a proton with the phosphate, would cause the phosphatidyl-choline to 'bob down' into the bilayer. Thereafter, the molecule might move through the bilayer to the other side in a flip-flop (suggested by Toyoshima & Thompson) or the H^+ and Cl^- might come off the molecule as HCl and escape to the other side of the membrane (as suggested by Robertson & Thompson, 1977).

The vertical movement of molecules need not be confined to the lipids or other low molecular weight substances. Proteins may also be subject to change in their relative hydrophilic/hydrophobic properties. Indeed, the first suggestion that the vertical movement of proteins in bilayers has functional significance was made by Blasie (1972) to explain the behaviour of rhodopsin after light absorption by its chromatophore, retinal. As we shall see, the mechanisms of rhodopsin and bacteriorhodopsin energy transfer after light trapping are not yet fully explained.

The extent of the movement of molecules at right angles to the plane of the membrane may be quite small, but it might have considerable effects.

72

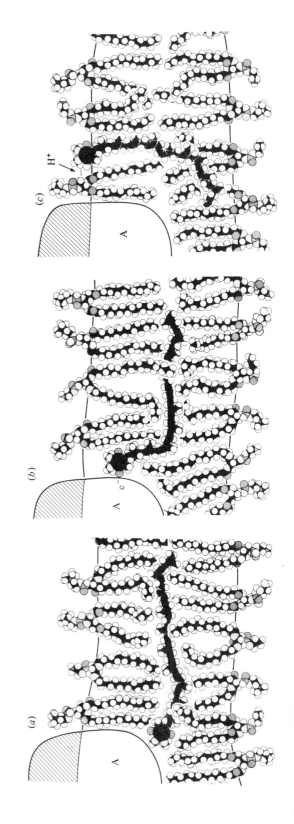

Fig. 4.6. The bobbing up and down of ubiquinone: (*a*) oxidised ubiquinone in the centre of the bilayer; (*b*) oxidised ubiquinone moves to pick up an electron from NADH oxidase (A); (*c*) the negatively charged quinol moves so that its head group enters the aqueous phase; (*d*) a proton attaches itself to the negative charge to form ubiquinol which bobs down in the bilayer; (*e*) the neutral ubiquinol returns to the lipophilic centre of the bilayer and diffuses to the right; (*f*) the reduced ubiquinone is oxidised again to ubiquinone by the cytochrome b/c_1 (B) complex accepting the electron and the released proton is picked up by an anion (Cl^-).

73

Fig. 4.6 (cont.)

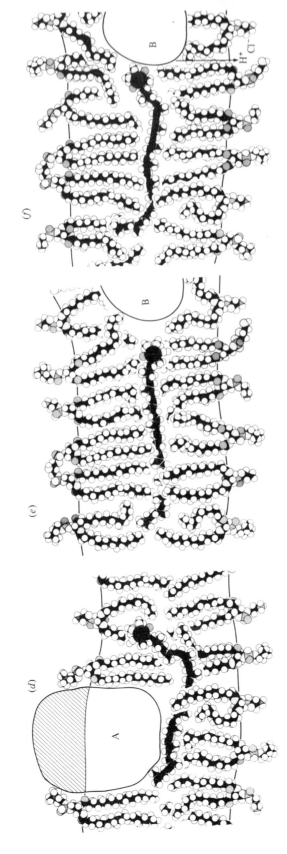

74

For instance, on becoming charged, the polar head of ubiquinone or plastoquinone may bring the molecule up by only about 0.4–0.5 nm, but that would be quite sufficient to remove the polar group from the non-aqueous environment. Similarly, when the protons are picked up and the charge decreased, the molecule will return the same short distance to have its head group back in the non-aqueous environment. These differences

Fig. 4.7. The bobbing up and down of phosphatidylcholine (PC): (a) PC combines with a $^{36}Cl^-$ at the positively charged choline and a proton at the negatively charged phosphate; (b) the decreased water-attracting properties of the uncharged polar group allow the phosphatidylcholine to sink into the bilayer; (c) the phosphatidylcholine in the non-aqueous environment loses its $^{36}Cl^-$ and H^+ which form $H^{36}Cl$; (d) the charged phosphatidylcholine returns to its place in the bilayer; the $^{36}Cl^-$ and H^+ exchange with H^+ and Cl^- on another phosphatidylcholine and (e) pass to the other side of the bilayer; (f) H^+ and $^{36}Cl^-$ have crossed the membrane by taking advantage of the phosphatidylcholine combination.

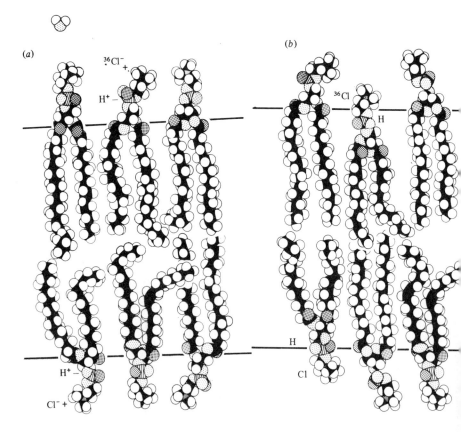

(a)

(b)

between reactivity in water and – such a short distance away – reactivity in a non-polar solvent may be of very great importance in increasing the range of types of reaction which membrane molecules can undergo in virtually two different types of solvent.

Fig. 4.7 (*cont.*)

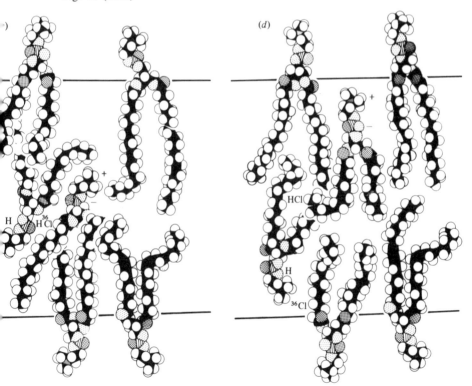

Fig. 4.7 (*cont.*)

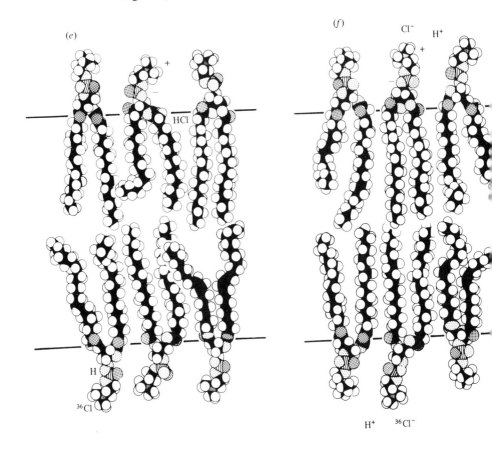

Conformational changes in proteins

Most enzyme reactions depend on conformational changes of varying degrees. As will be expected, such conformational changes occur during reactions in membrane-bound proteins and would be no different from those in other enzyme reactions, if it were not for the close proximity of the aqueous and non-aqueous membrane phases. This means that, in theory, an enzyme reacting with a substrate in the aqueous phase can undergo a conformational change in which the product is released in the non-aqueous phase. If, now, the two sides of the bilayer have different permeability to that product, it may be prevented from returning to the side of the substrate and leak only to the far side of the membrane. Thus, the membrane has a functional 'sidedness' – not a happy word but one in common use for implying that what happens is different on the two sides of the bilayer – and the result is a vectorial change. The phenomenon in which a substance, picked up by a membrane molecule on one side, is brought into the membrane and, when released, cannot return as fast as it can go to the other side of the membrane, must be very common. It is certainly the basis of the translocation of protons by the bacteriorhodopsin molecule though the details of the mechanism are obscure. It is also probably the basis for the 'gating' mechanism where, as we shall see, specific channels for Na^+ or K^+ may be opened or shut in nerve membranes. It also occurs in the ion-translocating ATPases (Ca^{2+}-ATPase and Na^+-K^+-ATPase).

Diffusion across bilayers: ionophores

The passage of smaller ions is by diffusion unless they are combined with, or carried by, one of the membrane-bound molecules. Entry into the membrane depends on

(1) ability to move among the polar groups to enter, and
(2) solubility in the hydrophobic phase.

We have discussed how ions can enter as hydrated ions with water molecules. Other substances acting as ionophores can be quite specific. The *cation-transport antibiotics* which are produced by some microorganisms are molecules which are able to penetrate membranes. Valinomycin is a cyclical peptide consisting of L-lactate, L-valine, D-hydroxyisovalerate and D-valine repeated three times round the ring (Fig. 4.8). It acts as a carrier for potassium because the metal ion is chelated by the oxygen atoms on the inside of the ring but, equally important is that the ring exposes only lipophilic groups on the outside. Thus valinomycin's solubility in the

78

membrane is considerable and a potassium ion inside it is moved through the lipophilic region of the membrane.

Fig. 4.8. Valinomycin, a cyclical peptide with L-lactate, L-valine, D-hydroxyisovalerate and D-valine repeated three times round the ring: (a) in surface view showing the way potassium is chelated by the oxygen atoms; (b) in side view, showing the predominantly lipophilic nature of the outside. The average diameter of the valinomycin ring is approximately equal to the length of one phospholipid hydrocarbon chain (c).

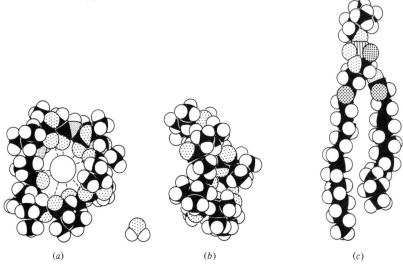

(a)   (b)   (c)

79

Lively, specialised molecular environment

Our slowness in recognising and explaining the cooperative movements of the specialised molecules in membranes has been an impediment to our understanding of membrane-bound reactions. The living membrane provides a molecular environment in which molecules range from those moving freely past each other to those stationary, in aggregates or attached firmly to other molecules like proteins. Molecules can be partly in water with hydrophilic reactions like hydrogen bonding and polar linkages and partly in a lipophilic environment of hydrophobic, uncharged molecules. Some parts of molecules can move from the aqueous to the lipophilic environment. Some, held tightly to proteins by hydrophobic forces, ionic linkages, or both, can change their bonding depending on charge (addition or loss of an electron or proton) and consequent vertical movements in the bilayer. Furthermore, the lipophilic portion of the bilayer is of sufficient dimensions to allow the cell to carry out some reactions in a non-polar solvent.

Altogether, the whole bilayer is a very lively but controlled system, perhaps best likened to a full ballet company with its decor and choreography. Our understanding of membrane reactions and processes will depend on our recognition of their molecular dances.

Suggested reading

Israelachvili, J. N., Marčelja, S. & Horn, R. G. (1980). Physical principles of membrane organisation. *Quarterly Review of Biophysics*, **13**, 121–200.

Lenaz, G. (1977). The role of lipids in the structure and function of membranes. *Subcellular Biochemistry*, **6**, 233–343.

5

Energy transduction – light-energy trapping

Now we have established the basic structure, organisation and movements in membranes, we can use this knowledge to attempt an understanding of *membrane-bound processes*, living processes which take place only in or on the surface of a membrane. In this chapter we shall consider three membranes in which the trapping of light and the conversion of the energy to other forms are completely dependent on membrane organisation.

The transformation of energy from one form to another in membranes, an almost universal phenomenon in living cells, is known as *energy transduction*. We shall see that the sources of energy may be in oxidation reactions or in an ion or proton concentration gradient. In energy transduction, the primary step is often a *charge separation*, i.e. an electron and a proton are separated and a potential difference is established across an insulating region which keeps the two apart. Though either a proton or an electron may move, the simplest way to understand what can happen is to consider a system in which a proton and electron, having been together as a hydrogen atom, are separated because the electron is picked up by an oxidising agent and dispersed through its electron orbits to a metal atom, e.g. iron. If such a molecule is buried in a non-polar region of the membrane, the proton left behind cannot move to the electron (Fig. 5.1) and will move to the point of lowest free energy. This usually means

Fig. 5.1. The principle of charge separation in membranes: (*a*) a carrier with hydrogen atom attached approaches the oxidising agent; (*b*) the carrier loses its electron to the oxidising agent and the proton is left; charge separation has taken place; (*c*) the electron passes to an acceptor, the proton moves out of the membrane; (*d*) the proton, now a hydroxonium ion in water and the electron remain separated by the insulation of the membrane.

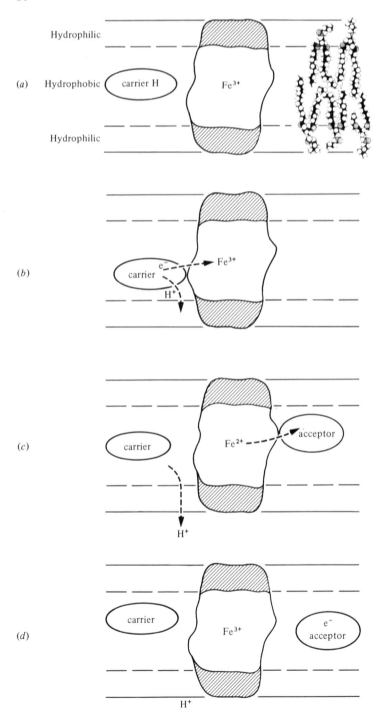

(a) Hydrophilic

Hydrophobic carrier H Fe^{3+}

Hydrophilic

(b) carrier e^- Fe^{3+}

H^+

(c) carrier Fe^{2+} acceptor

H^+

(d) carrier Fe^{3+} e^- acceptor

H^+

extrusion from the non-polar or insulating regions of the membrane to enter water and become a hydrogen (hydroxonium) ion. Meanwhile the electron may have been passed to another compound or (as with respiration) to oxygen and hydroxyl ions will be formed. If now the whole membrane with its insulating properties keeps hydrogen and hydroxyl ions apart, one side of the membrane goes acid and the other side goes alkaline. Various secondary adjustments can be expected, the simplest being on the redistribution of other ions across the membrane. More complicated consequences can follow the effect of such a gradient upon the membrane-bound enzymes and their substrates. The effects of charge separation on the subsequent distribution of ions and on reactions in membranes were the basis of Peter Mitchell's *chemiosmotic hypothesis*, first published in 1961 (Mitchell, 1966). This hypothesis was so different from the way that most leading biochemists thought, twenty years ago, that it met with considerable resistance. Williams (1961, 1962) was an exception, though his interpretation diverged from that of Mitchell. However, it soon became clear that Mitchell's explanation did so much to elucidate so many cell processes that biochemists, especially the younger ones with new ideas to formulate, accepted it. Its undoubted success in advancing science led to Mitchell's award of the Nobel Prize for Chemistry in 1978.

I shall discuss three light-trapping membranes before going on to other energy transductions. These membranes are responsible for life on earth as we know it today. They are also helpful in illustrating several different ways in which, after the energy is trapped, other kinds of energy transduction can occur. The first, the cell membrane of the photosynthetic bacterium, *Halobacterium halobium*, which flourishes in saturated salt solution, sets up a charge separation in light resulting in an extrusion of protons; the consequent gradient is used by the cell to make ATP for its synthetic reactions, or to swap ions around between inside and outside. The second is the retinal membrane of animal eyes in which the energy trapped sets up some kind of gradient which affects other membranes in the vicinity, leading to a signal passing along the nerve membrane. It can be regarded as a sequence of electrical events. In the third light-trapping membrane, thylakoid of the chloroplast, charge separation is followed by a series of events with membrane-bound molecules until ATP is formed by the proton gradient and nicotinamide adenine dinucleotide phosphate ($NADP^+$) is reduced by the passage of two electrons, picks up two protons and becomes NADPH. ATP and NADPH are made on the outside of the membrane. ATP carries its phosphorylating ability and NADPH its reducing ability to the other reactions of photosynthesis, some in the chloroplast and some in other parts of the cell.

Halobacterium: light-trapping, making ATP

In *Halobacterium halobium* the light-energy operates the bacteriorhodopsin, the retinal-containing protein, as a proton pump, i.e. it takes protons from inside the cell to the exterior and makes a potential difference between the two sides of the membrane; this difference allows for other processes essential to the life of the cell and is maintained until used for other purposes such as ATP formation, amino-acid transport or ion movement.

The cell membrane of *Halobacterium* occurs inside a cell wall which consists of protein, lipid and a small amount of carbohydrate (Fig. 5.2*a*). The purple part of the cell membrane is coloured by the bacteriorhodopsin and occurs in specialised patches responsible for light-trapping (Fig. 5.2*b*). The other (red) parts of the membrane are responsible for all other cell membrane functions including respiration in the presence of oxygen (Fig. 5.2*c*).

The amount of purple membrane is controlled by the cell and, if the oxygen supply drops below a critical level, part of the red membrane is replaced by purple which may increase to occupy up to 50% (wt/wt) of the cell membrane. The bacteriorhodopsin (m.w. 26 000) protein occurring as a precise molecular array in the purple patch accounts for 75% (wt/wt)

Fig. 5.2. A diagram of a *Halobacterium* cell: (*a*) the cell wall; (*b*) a patch of purple membrane; (*c*) the red membrane; (*d*) an enlargement of a section of the membrane showing the path of H^+ to the F_0 of the ATPase and synthesis of ATP in the F_1.

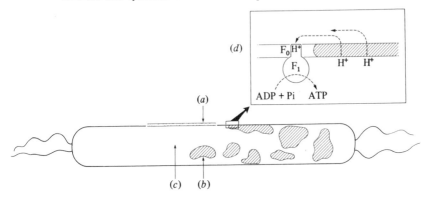

of that membrane, accompanied by 25% (wt/wt) of lipid or about 10 lipid molecules per bacteriorhodopsin molecule. This contrasts with the red membrane which contains many different proteins but no bacteriorhodopsin. The lipids in the purple membrane are phosphatidylglycerophosphate (52%), phosphatidylglycerol (4%), squalenes, which are C_{30} neutral lipids each made up of 6 isoprenoid units, some other neutral lipids (about 9%), two sulphur-containing lipids, phosphatidylglycerosulphate (5%) and a glycolipid sulphate (10%). A glycolipid (triglycosyldiether) makes up about 19%. The phospholipid and neutral lipid composition is almost identical in both the red membrane and the purple membrane but it may be very important that the sulphur lipids and the triglycosyldiether are not found in the red fraction. The lipids are all believed to consist of dihydrophytol chains (Fig. 5.3) linked to glycerol by ether linkages. The dihydrophytol chain is saturated but branched; its methyl groups probably prevent the close packing of the hydrocarbons and favour fluidity.

We have described the remarkable molecule of bacteriorhodopsin with its seven rods of α-helical polypeptides, each about 4 nm long and about 1 nm apart (Fig. 2.16). Attached to a lysine is a molecule of retinal (which is vitamin A, except that the terminal group is an aldehyde instead of an alcohol), the same chromatophore as occurs in the rhodopsin of eyes (Fig. 5.4a).

In the purple patches, the bacteriorhodopsin molecules are arranged in trimers. The absorption of light occurs in the retinal chromophore and is followed by a series of, now well understood, reactions. The retinal,

Fig. 5.3. The hydrocarbon chains of the lipids of bacteriorhodopsin consist of dihydrophytol.

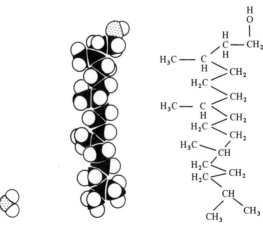

attached as a Schiff base to a lysine (Fig. 5.4b), which is either number 41 (Brigden & Walker, 1976; Engelman *et al.*, 1980), or number 216 (Mullen, Johnson & Akhtar, 1981), in the polypeptide chain (see Fig. 2.16), near the inner surface of the membrane, is protonated in the dark. On absorbing light it is converted from the 13-*cis* isomer to the all-*trans* isomer (Fig. 5.5). This change, which occurs in about 10 ps, results in an alteration in the geometry of the molecule, but the protonation is apparently not altered immediately. Subsequently, two other isomers can be detected by absorption bands; the first appears with a half-time of about 2 μs and the second with a half-time of about 40 μs. Then, with a half-time about 5 ms, two other

Fig. 5.4. (*a*) Retinal; (*b*) retinal attached as a Schiff base to the lysine side chain of the protein of bacteriorhodopsin.

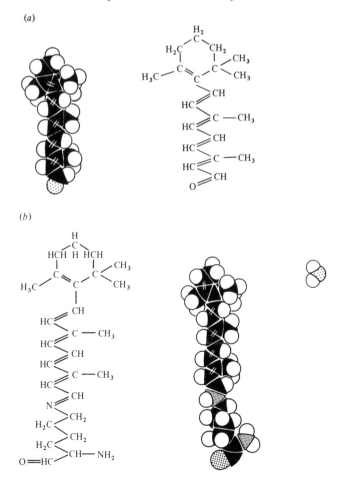

(*a*)

(*b*)

things happen: first, the proton comes off the Schiff base and second, the molecule undergoes at least one more spectroscopic change before reverting to the first member of the cycle by picking up another proton. The first proton which had been picked up on the inside of the cell membrane is now on the outside. These reactions depend on the presence of adequate lipids in the membrane.

A major gap in our knowledge of this deceptively simple system is how the proton, which has been picked up on the Schiff base very near the inside of the membrane, appears on the outside. How does the change in the protein of retinal allow the movement of that proton across the cell membrane? The proton release to the solution, based on work with resonance Raman spectra, seems to be correlated in time with the later changes in the retinal protein after light absorption. Whatever the details of the mechanism, the fact remains that the proton, picked up on the inside of the membrane, is prevented from going back and is moved across to the outside. Applying our knowledge of molecule movements in membranes, there are three possibilities:

(1) there has been sufficient movement in the protein-retinal molecule for the proton to be well across the lipophilic layer (about 3.5 nm) before it is released from the Schiff base. This could conceivably be due to a considerable vertical movement of the whole protein

Fig. 5.5. Retinal shown in the all-*trans* form (*a*) which it assumes after absorbing light and in the 13-*cis* form (*b*); note the difference in the space occupied by the two forms.

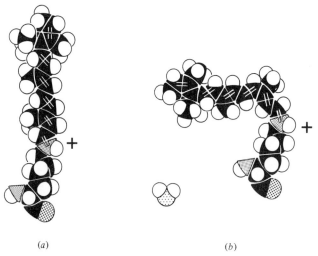

(a)   (b)

similar to that suggested by Blasie (1972) for rhodopsin in eyes or to a large conformational change;

(2) the proton is released into the lipophilic region by being combined with a lipophilic membrane molecule which then bobs out into the aqueous phase to lose its proton;

(3) the proton may come off the Schiff base near the inside of the membrane but then move through a proton-conducting channel, perhaps in the protein itself.

Engelman *et al.* (1980) have pointed out that the presence of a few charges within the core of the protein molecule might provide a network through which protons could jump – in a fashion analogous to the movement of protons in a channel ionophore of the gramicidin type (Chapter 1 and 7).

The effect of the pump would be cancelled if the proton could diffuse back into the cell, so the membrane must be impermeable to protons. This high resistance to the back diffusion would be due to the properties of the lipid bilayer. For example, the proton, even if attached to a small mobile lipophilic molecule, may not be able to escape to the cellular side of the membrane through closely packed hydrocarbon chains. We have seen that head groups of glycolipids, here represented by triglycosyldiether, can hold together by hydrogen bonds and, as a result, can cause the adjacent parts of the hydrocarbon chains to hold more tightly together, thus increasing the resistance to diffusion between them. So the membrane could allow diffusion to the outside while preventing back diffusion to the inside. However, we must know more about the proton pathway in the purple membrane before we can decide exactly what happens.

Whatever the mechanism which results in the establishment of a gradient of protons across the membrane, the protons then act on the ATPase which occurs in the cell membrane, resulting in ADP and P_i within the cell being converted to ATP (Fig. 5.3*d*). The ATPase of *Halobacterium* seems to be similar in general form and in principles of action to the ATPases which were described in Chapter 2.

The bacteriorhodopsin system lends itself to experiments in reconstituted vesicles. If vesicles are made from phosphatidylcholine, bacteriorhodopsin will move into the bilayer and, in the presence of light, will move protons from the external solution to the *interior* of the vesicle. Note that the polarity is different from that of the natural membrane which moves protons to the *exterior*; ATPases from several sources are compatible with the phosphatidylcholine and bacteriorhodopsin in such vesicles (Fig. 5.6). If such an ATPase is put into the bilayer, some will orientate with the F_1 towards the outside of the membrane. If ADP and inorganic phosphate are supplied to the outside and the system is illuminated, the bacterio-rhodopsin pumps protons to the interior of the vesicle, they enter the F_0 of the ATPase and bring about the synthesis of ATP from the ADP and inorganic phosphate in the F_1 on the outside of the vesicle.

How the protons, which have been built up in considerable concentration on the outside of the cell or the inside of the vesicles, enter the lipophilic tail piece (F_0) of the ATPase is unknown. Two broad possibilities can be suggested: first, that the F_0 has a proton-conducting channel capable of picking up the proton from the aqueous phase and transferring it to the seat of the ADP–P_i reaction in the F_1; second, that some of the protons liberated by the bacteriorhodopsin may move laterally in the bilayer on a small mobile lipophilic molecule and enter the F_0 in a lipophilic environment. There is also the possibility that protons in the aqueous phase may be re-introduced to the non-aqueous phase by a mechanism such as that described for phosphatidylcholine, hydrogen and chloride in Chapter 4, i.e. a proton may attach to a negatively charged lipid, decrease its hydrophilicity and move back into lipophilic centre of the membrane.

Fig. 5.6. A diagram of a vesicle containing (a) ATPases and (b) patches of bacteriorhodopsin. The dotted lines show the H^+ pathways. ATP is synthesised on the outside.

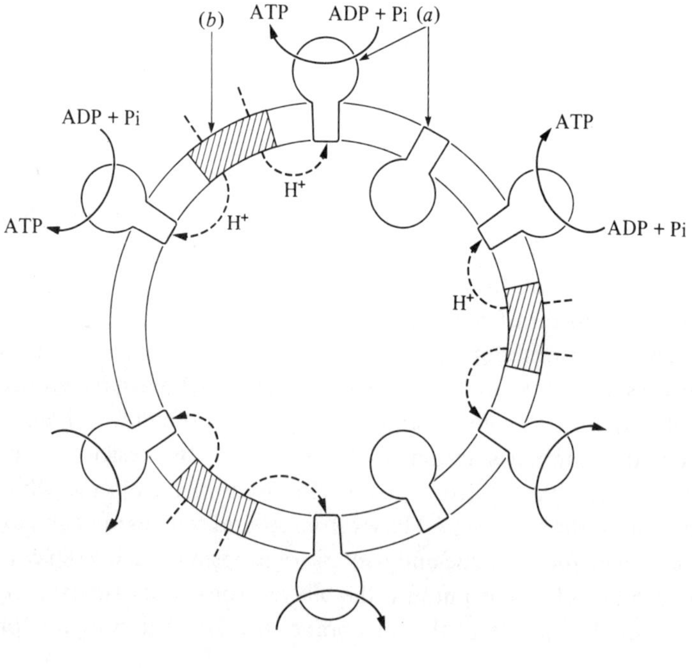

The possibility that protons can be put into the non-aqueous phase of the membrane by bacteriorhodopsin is plausible, as shown by recent experiments by Post and Young in my laboratory. An emulsion of octane in sodium chloride or sodium nitrate can be made with a mixture of phosphatidylcholine and bacteriorhodopsin as emulsifying agent. The bacteriorhodopsin is situated on the emulsion droplets in the interface between octane and water. When the emulsion is illuminated protons are picked up from the aqueous phase, presumably entering the non-polar octane as the undissociated hydrochloric or nitric acid, depending on which anion is present. In light, a steady state concentration of protons results when the uptake is equal to the back diffusion. In darkness, the protons return to the external solution. Some such uptake into the lipophilic region of the bilayer is possible but the exact mechanism of proton entry into ATPases requires more understanding of their protein composition and behaviour and, in the meantime, must remain speculative.

This relatively simple light-trapping system of bacteriorhodopsin, given the appropriate orientation in a membrane, converts the light energy absorbed, first into a charge separation, after which protons moved can establish the necessary gradient to synthesise ATP; in the natural system, protons moved to the outside are effective in the ATPase which has its F_1 to the inside so the synthesis is within the cell. Alternatively, the gradient of protons may be used by the cell to drive potassium ion uptake or amino-acid transport (Chapter 8).

Eyes: light-trapping, potential difference, nerve signal

Light-trapping mechanisms in eyes are somewhat similar, in their light reactions, to the bacteriorhodopsin system but the consequences are quite different and much more complicated. The energy of the light absorbed alters the distribution of ions across the surface membrane of the receptor cell and thereby changes the membrane's electrical potential. The potential difference in the dark – the *resting potential* – is maintained by the cell pumping out Na^+ and pumping in K^+ (see Chapter 8). Once initiated by the light reaction, the change in potential travels along the cell membrane like nerve impulses, changing the potential temporarily in successive spots until it reaches the end. Thus the initial signal, due to the light, is amplified by a factor of 1000–2000 due to successive changes in the existing potential (see Chapter 9).

Light-trapping membranes of eyes resemble *Halobacterium* membranes because they contain a pigmented protein, rhodopsin, which is very similar to bacteriorhodopsin. About half the dry weight of the light-trapping membranes in eyes is lipid and about 40% is protein, of which 80–90% is rhodopsin. Reduction in the amount of lipids results in reduced light absorption. Rhodopsin, a glycoprotein which spans the membrane, has a molecular weight of 35000–38000 with a polypeptide chain of 350 amino acids; the amino acid sequence is not fully known. As in bacteriorhodopsin, rhodopsin has retinal linked to a lysine by a Schiff base near the surface of the membrane. The retinal changes from an 11-*cis* form unlike that of bacteriorhodopsin which has a 13-*cis* form, to all-*trans* form, on absorption of light. In both invertebrate and vertebrate eyes, the rhodopsin light absorption is followed by a succession of changes, seen spectroscopically, and somewhat similar to those of bacteriorhodopsin.

In invertebrates, the pigment rhodopsin is part of a highly folded cell membrane (see Chapter 1, Fig. 1.3), so it has a large surface area for photon capture. A resting potential across the membrane is maintained in the dark with K^+ in high concentration inside and Na^+ in low concentration. This difference is maintained by an ATP-dependent Na^+/K^+ pump of the type to be described in Chapter 8. In an arthropod, for example, K^+ inside may be about 20 times that outside while Na^+ may be about one tenth and Ca^{2+} may be about one thousandth. The potential difference is mainly due to the K^+ gradient with 50–70 mV (inside negative) in the dark. When the light is absorbed a transient increase in membrane conductance occurs, ion currents flow across the membrane as the ion gradients and potential are reduced. The Na^+/K^+ pump restores them again later. How the change in conductance is caused by changes in the retinal on rhodopsin is unknown; it seems probable that the first effect is a charge separation in the membrane, if the rhodopsin behaves like bacteriorhodopsin. However, whatever the mechanism, this change in conductance initiated at one point – even one photon absorbed by rhodopsin will suffice – spreads over the whole cell membrane. Finally, the change in potential transmits a signal to a nerve cell which, in the manner described in Chapter 9, carries the electrical impulse to the brain. In the light-receptor cell, the light energy triggers an electrical response but the membrane potential, maintained by respiration via ATP, is responsible for the amplification of the signal.

In the vertebrate eye, the pigment is not in the surface membrane but is contained in stacks of discs in cells which, according to shape, are known as retinal rods or cones (Fig. 1.2, Chapter 1). The membranes of the discs are separated from the cell membrane by 0.02–0.04 μ. A retinal rod in section is shown diagrammatically in Fig. 5.7. In the rod, light absorption by rhodopsin in the discs ultimately causes a decrease in cell membrane conductance. Since the disc membranes are some distance from the cell

91

Fig. 5.7. A diagram of a longitudinal section through a retinal rod: (*a*)
outer segment containing disc stack (DS), separate but being formed
from the cell membrane (DF); (*b*) inner segment containing
mitochondria (M), Golgi body (GB), rough endoplasmic reticulum
(RER), nucleus (N); (*c*) end of cell making synaptic connection with
nerve.

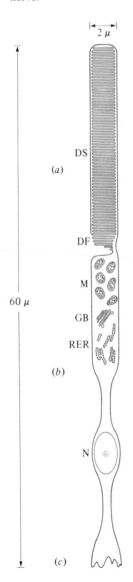

membrane in which the decreased conductance occurs, there must be some connecting mechanism, presumably in the form of a soluble transmitter molecule or ion. Exactly what happens in the charge separation within the disc membrane is not known. The current hypothesis is that Ca^{2+}, pumped into the discs in the dark, and released quickly through the disc membrane after the light/rhodopsin reaction, diffuses to the cell membrane. There, it is suggested, it closes channels through which sodium (in high concentration outside) has been diffusing inwards, decreases the membrane conductance and thus increases the polarisation of the membrane. This increase in potential is sufficient to trigger an electrical impulse which moves to the end of the rod cell and there activates the connected nerve.

Though many details of the system (e.g. how the Ca^{2+} is pumped into the discs, how the light/rhodopsin reaction results in its release) await further research (Bonting & Daemen, 1977; Hubbell & Bownds, 1979), the eye provides a beautiful example of an integrated system of membrane-dependent reactions. Thus, introducing the light-trapping membranes of eyes at this stage allows me to make three important points about membranes, which would not be apparent from bacteriorhodopsin:

(1) The light may provide only the triggering energy for the membrane changes which follow.

(2) The triggered mechanism can result in an amplification of several orders of magnitude. This is from energy stored as a membrane potential established as a result of metabolic activity providing ATP for a specific ion pump.

(3) Changes in one membrane (that of the discs in the vertebrate retinal rod) can be transferred, using a soluble transmitter (in this case possibly Ca^{2+}) to another membrane some small distance away.

Photosynthesis: light-trapping, ATP formation, reducing power

In photosynthesis we see two charge separations in sequence, each requiring a quantum of light to make up the necessary amount of energy. The principles are illustrated in Fig. 5.8 and we shall attempt to understand the various membrane reactions which are involved. All the reactions take place in the *thylakoid membranes* of the chloroplast (Anderson, 1975; Clayton, 1981).

The appearance of the thylakoid membrane is seen in the electron micrograph of Fig. 5.9. Some thylakoid membranes are arranged in stacks, called *grana*, and are easily distinguished from the other thylakoids which cross the *stroma* of the chloroplast between the grana (Figs. 5.9, 5.10*a*).

93

As we shall see, there is evidence that these membranes have different functions. A thylakoid consists of two bilayers with an intra-thylakoid space. Each bilayer consists of proteins and lipids in about equal amounts. The lipids are about 80% galacto- or digalactolipids, which may be

Fig. 5.8. Principles of two charge separations each dependent on light absorption: (a) the first light (arrow) reaction results in a charge separation when an electron moves away from a proton in photosystem II; (b) the electron is picked up by an acceptor which also acquires a proton and moves to become the second donor to photosystem I; (c) the second light (arrow) reaction causes the second charge separation.

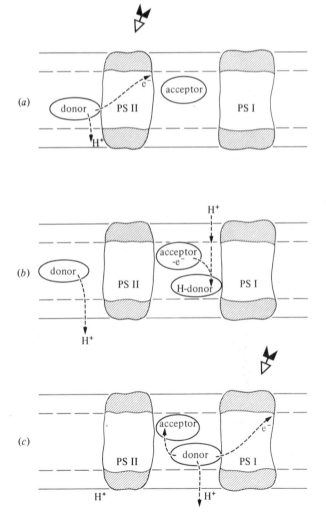

94

Fig. 5.9. Electron micrograph of a chloroplast showing the parallel thylakoid membranes, in some places arranged in a stack or granum (G). Also in the chloroplast are starch grains (SG), plastoglobuli (PG). Outside the chloroplast are cell wall (CW), plasmalemma (PL), mitochondria (M) and vacuole (V). (Micrograph: J. M. Bain.)

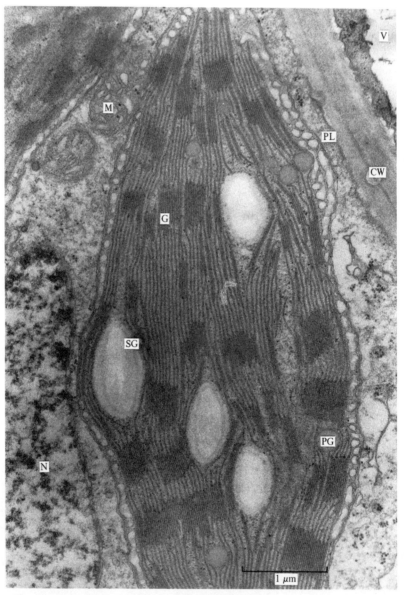

significant in the function of this light-trapping membrane. In a single thylakoid of the granum stack there is a large number of pigment molecules consisting of chlorophyll *a* and *b* and carotenoids. The charge separations are brought about by a number of electron carriers, referred to collectively as an electron transport chain (Fig. 5.10*b*). There are about 500 pigment molecules for each electron transport chain. Most of these are the light-harvesting pigments attached to proteins, known as LHCP or 'antennae pigments' – they collect the light signals just like a radio antenna collects radio signals. When a pigment molecule is 'excited' by a quantum of light, an outer shell orbital electron is raised to a higher energy level and this excitation energy may be passed from one pigment to another and so reach the specialised chlorophylls which are responsible for the two charge separations and, as shown in Fig. 5.8, one side of the membrane goes negative and the other side goes positive.

The photoactive systems are associated with two different chlorophyll *a*-protein complexes: photosystem I which absorbs light of wavelength up to 700 nm, and photosystem II which absorbs light up to a wavelength of 680 nm, and the two acting together, in series, bring about the two charge separations (Fig. 5.11). On receiving light or excitation energy, photo-

Fig. 5.10. (*a*) Thylakoids in grana (G) and crossing the stroma (S); (*b*) thylakoids enlarged to show the difference in distribution of photosystem I (black square) and photosystem II (hatched square). Note that the ATPase (F_1, F_0) does not occur in the granum. The electron transport chains are in the membranes.

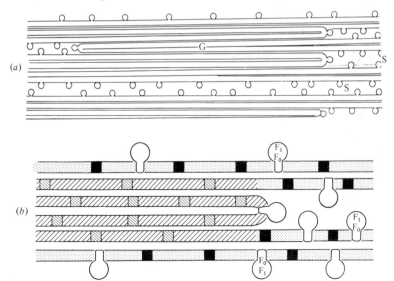

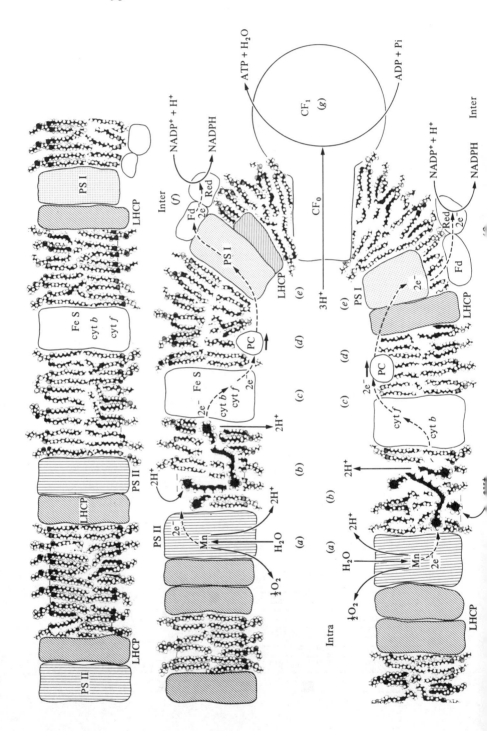

system II, in conjunction with an unknown enzyme system in which manganese and chloride ions are involved, takes electrons from water (Fig. 5.11*a*). Thus the electrons pass to the chlorophyll *a*, already oxidised by light, towards the outside of the membrane, oxygen is evolved and protons are released on the inside of the membrane. Charge separation within the membrane is established in only 2 ns. The electrons pass from these specialised chlorophylls, following excitation by light, to the next member of the electron transport chain, plastoquinone, towards the outside of the membrane (Fig. 5.11*b*). Then, for each electron it receives, plastoquinone picks up a proton on the outside of the thylakoid and brings it into the membrane. When the reduced plastoquinone loses an electron to the next electron carrier in the chain, which is a complex of cytochromes *b*, *f*, and FeS proteins, the proton is left unpaired and passes to the intra-thylakoid space (*c*). The electron passes through the complex to plastocyanin, a small water soluble molecule, which adsorbs to the membrane in the intra-thylakoid space (*d*).

Meanwhile, the other chlorophyll in photosystem I of the electron transport chain is ready to receive energy and, after the absorption of one photon, becomes oxidised and can accept an electron from the plastocyanin (*e*). This electron is then donated to ferredoxin. Two molecules of reduced ferredoxin, each having received an electron, react with the enzyme ferredoxin-$NADP^+$-reductase which in turn reduces nicotinamide adenine nucleotide phosphate ($NADP^+$) on the outside of the thylakoid to form NADPH, the reducing catalyst of photosynthetic cells (*f*). Note that the negative charge moves from the inside to the outside of the thylakoid membrane. The NADPH so formed is released into the stroma of the chloroplast where it goes about its activities in the chemical fixation of carbon dioxide. The net result of the second light-absorbing act, therefore,

---

Fig. 5.11. The hypothetical sequence of photochemical reactions, electron transport and ATP synthesis in thylakoids: (*a*) photosystem II, PS II, reacts with water and light to give 2 electrons, 2 protons (in the intrathylakoid space, Intra, of the granum) and oxygen; (*b*) the 2 electrons pass to oxidised plastoquinone which picks up 2 protons from the interthylakoid space, Inter; (*c*) the 2 electrons pass on to cytochrome *b–f* complex and 2 protons pass to the intrathylakoid space; (*d*) electrons pass to the carrier, plastocyanin, PC, which passes them to the light-activated photosystem I, PS I at (*e*); (*f*) electrons from PS I pass to ferredoxin. Fd, which in conjunction with $NADP^+$ reductase, Red, reduces $NADP^+$ and H^+ to NADPH; (*g*) protons from the intrathylakoid space enter the ATPase to react with ADP and P_i to form ATP and H_2O in the interthylakoid space. Light harvesting chlorophyll proteins, LHCP, are involved with both PS I and PS II.

is a second very rapid charge separation across this remarkably organised membrane.

There are further complications to the story just presented, some of which are only now being elucidated. It would be easy to account for the movements of the electrons between photosystem II and photosystem I if they lay in close proximity with the electron carriers between them but they do not. The grana and stroma thylakoid membranes do not have the same structure nor the same function. Freeze-fracture electron microscopy reveals a marked difference between the sizes of particles occurring on the thylakoids in the grana partitions and of those on the thylakoids in the stroma. A combination of immunological studies and proteolytic experiments has shown that the ferredoxin-$NADP^+$-reductase, which passes the electrons from photosystem I to the $NADP^+$, is located only on the exposed thylakoid membranes in contact with the stroma, whereas the membranes within the grana stacks contain mostly photosystem II (Andersson & Anderson, 1980). This configuration of thylakoid membranes means that there must be some lateral shuttle of electrons on, or in, the membrane to take them from photosystem II to photosystem I. The electrons can be carried to the cytochrome/iron-sulphur complex by the lipid-soluble plastoquinone, which diffuses very quickly in the lateral plane of the membrane. Transfer of electrons from the cytochrome/iron-sulphur complex might be by the water-soluble plastocyanin.

The other non-homogeneity in thylakoid membranes poses another problem of movement. We have seen earlier that, at two points, the splitting of water and the oxidation of the plastoquinone, protons go to the inside of the thylakoid membrane, i.e. into the intra-thylakoid space. Such proton gradient across a membrane can be used to bring about the synthesis of ATP in a membrane-bound ATPase (Fig. 5.11g). Freeze-etching studies and antibody labelling showed that the chloroplast ATPase was located only on the stroma-exposed surfaces of thylakoids, not on surfaces adjoining other stacked thylakoids (Fig. 5.10a, b). Thus, the protons must travel quite a distance to the site of the ATP synthesis where they enter the CF_0 and move into the CF_1 which, in the presence of ADP and inorganic phosphate in the stroma, synthesises ATP. How they move is not definitely known; the commonly accepted view is that they move in the intrathylakoid water but the mobile plastoquinone, as we have seen, may carry those, picked up from the outside of the thylakoid, for part of the distance in the hydrophobic portion of the membrane.

The properties of the thylakoid membrane are likely to be governed by the relatively high quantity of galactolipids (80% of the total diacyl lipid).

One is a galactose linked to glycerol with two hydrocarbon tails, and the other is two galactoses, glycerol and two hydrocarbon tails (Bishop *et al.*, 1980). The hydrocarbon tails of both mostly have 18 carbon atoms and three double bonds, the first one at the ninth carbon atom from the glycerol. These two lipids are capable of quite strong hydrogen bonding in their head groups. Consequently, as we have seen, the first eight $-CH_2$ groups from the head group before the first double bond at the ninth carbon atom, will tend to be held tightly so that their hydrophobic forces will overcome their kinetic movement and they would constitute a barrier of high resistance to diffusion of molecules tending to escape from the membrane. By contrast, the nine carbon atoms below the double bond would maintain the highly fluid state of the centre of the bilayer and provide the ideal environment for rapid lateral diffusion of a lipid-soluble substance such as plastoquinone.

Much remains to be done experimentally before all the molecular details of light-trapping membranes, with their wealth of lively reactions, are properly understood. Already, they clearly illustrate the essential membrane activities which would be impossible without the molecular structures, movements and reactions which characterise the realm of the bilayer. They have been introduced at this stage to stress the absolute dependence of these characteristic living reactions on the organisation of the lipid–protein–pigment bilayers.

Suggested reading

Clayton, R. K. (1981). *Photosynthesis: Physical Mechanisms and Chemical Patterns*. Cambridge: Cambridge University Press.

Davson, H. (1980). *Physiology of the Eye*, 4th edn. Edinburgh, London & New York: Churchill Livingstone.

Henderson, R. (1977). The purple membrane from *Halobacterium halobium*. *Annual Reviews of Biophysics and Bioengineering*, **6**, 87–109.

Hinkle, P. C. & McCarty, R. E. (1978). How cells make ATP. *Scientific American*, **238**, No. 3 (March), 104–23.

6

Energy transduction – oxidation-reduction energy

The thylakoid light-trapping membranes, like those of the algae and the photosynthetic bacteria, start reactions in which the protons and electrons, liberated from water, are transferred to carbon dioxide which becomes carbohydrate. Hydrogen atoms on the carbon-containing molecules remain a source of potential energy, to be freed again in respiration when they come off the carbon and join oxygen to form water. The process of respiration liberates energy as the electrons and protons are shuffled round, sometimes together, sometimes apart, but always in controlled reactions so that small portions of the total energy are made available at each step. Much of the energy is transferred into ATP, that important energy currency of all cells, which in turn makes its energy available again when hydrolysed back to ADP and inorganic phosphate. The control mechanisms for the redistribution of the energy in respiration are mostly in membranes. Since the mechanism depends on electron-transport chains which consist of enzymes that are alternately reduced and oxidised as the electrons pass through, we can refer to the energy as *oxidation-reduction energy* and discuss the associated energy transduction. It was due largely to his interest in these types of membranes that Peter Mitchell (1961) first published his chemiosmotic hypothesis.

I shall describe two examples of oxidation-reduction membranes, both of which convert the bond energy of molecules containing carbon, hydrogen and oxygen to a charge separation and then to the phosphate-bond energy of ATP. The first is the well-known system of the cytoplasmic membrane of the bacterium *Escherichia coli* (Fig. 6.1*a*), the second, that of the inner membrane of mitochondria, the ubiquitous organelles of eukaryotic organisms.

100

Cell membrane of *E. coli*

Charge separation: proton secretion

The cytoplasmic membrane of *E. coli*, which lies inside the periplasmic space separating it from the outer membrane and cell wall (Lugtenberg, 1981) is a bilayer containing many enzymes (Fig. 6.1*b*). For the moment we need to concentrate on the occurrence and properties of the electron-transport system which takes electrons, previously on substrate

Fig. 6.1. (*a*) The *E. coli* cell showing the flagella. (*b*) The surface layers of the *E. coli* cell. The cell wall (CW) which is freely permeable to substances in solution as is the outer membrane (OM) which is a bilayer of lipid polysaccharides in the outer layer (LPS) and probably phosphatidylethanolamine (PE) in the inner layer. Structural proteins (SP) separate the two membranes in the periplasmic space (PERI). In the cytoplasmic membrane (CM), protons move in with the amino acids (AA) and sugars (SS) and exchange with the Ca^{2+} and the Na^+ which are extruded. The periplasmic space is acid and the cytoplasm (CYT) is alkaline.

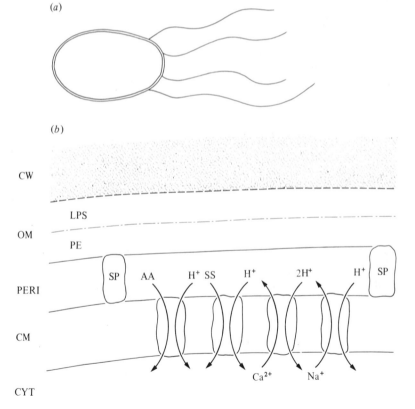

molecules or on water, and transfers them to oxygen, simultaneously secreting protons from the cell (Fig. 6.2). As we shall see, the proton gradient so established can be used for various purposes by other enzymes in the same membrane (Harold, 1977; Hinkel & McCarty, 1978). Most electrons are carried from the substrate-oxidising enzymes within the cell by nicotinamide adenine dinucleotide (NADH), a substance closely related to the NADPH which we saw accepting electrons in the thylakoids (Fig. 5.11). The loss of two electrons from the NADH results in its being oxidised to NAD^+. The electrons enter the membrane via a flavin adenine dinucleotide (FAD) which is attached to a membrane-bound protein. The proton that was attached to NADH, and another proton, are transported by the FAD protein to the outside of the membrane. Some compounds, e.g. lactate and succinate, can interact directly with the FAD of the membrane, losing two protons and two electrons into the FAD protein. The electrons pass through the FAD protein to the next members of the electron transport chain which are two nonhaem-iron sulphur proteins. The molecular mechanism which allows these electrons to pass through the FAD to iron sulphur protein is not understood. The electrons and protons have taken different paths in the membrane-bound system, because there are molecular electron conductors within the membrane insulation, resulting in a charge separation. The protons which have been liberated towards the *outside* of the cell membrane pass to the exterior

Fig. 6.2. The energy transduction of the *E. coli* cytoplasmic membrane: (*a*) the first charge separation in FAD; (*b*) electron transport to ubiquinone; (*c*) ubiquinone picks up protons from inside; note that it is pictured as picking up electrons at (1), picking up protons at (2) and losing electrons and protons at (3); (*d*) the second charge separation and reduction of oxygen; (*e*) the protons from outside make ATP on inside.

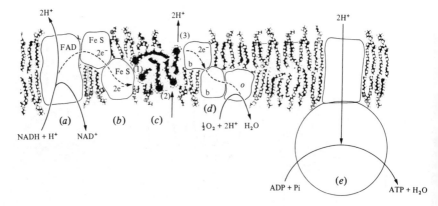

through the periplasmic space and the very permeable outer membrane and cell wall. Meanwhile, the iron sulphur protein complex transports the electrons back towards the *inside* of the membrane. There, they reduce either a single molecule of ubiquinone or two molecules in succession. The negatively charged ubiquinone molecules pick up one proton per electron from the aqueous phase inside the cell. The ubiquinone is then oxidised by passing its electrons to cytochrome *b* and the protons, which have been left, move to the *outside* of the membrane, pass into the periplasmic space and then out of the cell through the outer membrane and wall. Meanwhile, the two electrons pass back through cytochrome *b* to the inside of the cytoplasmic membrane where they are handed over to cytochrome *o* which catalyses the formation of water with oxygen. Thus, for each pair of electrons passing through the electron-transport chain, one oxygen atom is reduced and four protons go to the outside – two of these protons are those originally picked up from the substrate molecules, and two are from water. The inside of the cell tends to go alkaline, the outside to go acid and this proton gradient can be used for several energy-requiring purposes. It must be stressed again that the whole process is made possible only by the membrane's organisation and three particular features, similar to those we have seen in the thylakoid, are essential:

(1)   the membrane's insulating capacity;

(2)   the correct orientation of the proteins to allow the conduction of electrons through the oxidation-reduction chain;

(3)   the mobility of the ubiquinone in the membrane to pick up protons and to bring them into the bilayer from the cytoplasmic solution.

The living activity of *E. coli* depends on the versatility of this lively membrane.

The H^+ gradient makes ATP

Situated in the membrane is an ATPase similar to that which has already been described (Fig. 6.2). The knob (F_1), or ATP synthesising region, is on the inside of the membrane with a total molecular weight of 360 000 to 390 000, and five different subunits in the ratio $3:3:1:1:1$. The lipophilic base piece (F_0) is in the membrane itself and, by a mechanism unknown, protons from the outside pass through the F_0 to the F_1 to catalyse the reaction in that region resulting in the formation of ATP from ADP and inorganic phosphate. Thus, the electron-transport chain has established the proton gradient necessary to convert the charge separation into phosphate bond energy used in innumerable reactions in the cell.

The H⁺ gradient, uptake, loss of ions and molecules

We shall be discussing mechanisms of transport in a later chapter. For present purposes it is enough to record that the cytoplasmic membrane can absorb most common amino acids and some sugars (e.g. lactose, glucose, galactose) at their respective sites in the membrane (Fig. 6.1b). In each molecular absorption, an accompanying proton moves from outside to inside. Both Na^+ and Ca^{2+} are secreted from *E. coli* by mechanisms which depend on an exchange of positive charges, one H^+ inwards for one Na^+ outwards and $2H^+$ inwards for one Ca^{2+} outwards.

H⁺ gradient powers the flagellum

The *E. coli* cells swim through their aqueous medium due to the rotation of a flagellum. This is a long structure – up to 3 times the length of the cell – protruding from a hook which rotates in a bearing within the cell wall. The result is that the flagellum lashes in a circle and screws the cell through the water (Fig. 6.3). The inner end of the flagellum which

Fig. 6.3. An impression of the essentials of the mechanism for flagellal movement. The ring of 16 proteins (PRc) in the cytoplasmic membrane rotates because of some reaction between protons and the stationary ring (PRw) of proteins in the outer membrane and cell wall. Through the wall, the rotating shaft (SH) of the flagellum turns in a stationary bearing (BE) in the wall and this turns the hook (H) outside the cell. Since the long length of the flagellum comes out of the hook at a right angle, the flagellum (F) moves through a circle and propels the cell through the water.

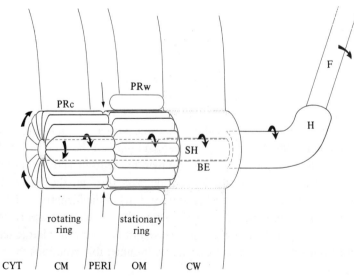

projects through the wall seems to lie in the cytoplasmic membrane region and to consist of a ring of 16 proteins. Another 16 proteins in the outer membrane remain stationary. The protein ring on the flagellum in the membrane appears to rotate when there is a gradient of protons from outside the cell inwards and this turns the flagellum (Berg, 1974, 1975; Harold, 1977). The analogy is to an electric motor. If one proton is required to move the mobile protein to the next fixed protein, like a kind of motor, it would require 16×16, i.e. 256 protons for each turn. There is still much to be learned about this translation of proton gradient energy into the energy of motion and about the changes in the proteins which bring about the rotation.

What an exciting idea, in principle, that this lively membrane has a motor dependent on positive current, especially as it has so often been said that Nature could not invent the wheel until man's ingenuity was brought into play!

The mitochondrial membranes

My own interest in charge separation goes back to the mid-1940s when I realised, with great personal satisfaction, that the separation of charge associated with electrons moving into cytochrome might be the basis of the mechanism of ion uptake by plant cells. I remember the Easter Saturday in 1945 when, sitting at a window of our house overlooking an Australian surf beach, I thought that my data for salt respiration and salt uptake should be quantitatively related if the hypothesis were tenable; on checking my experimental results, I found it was indeed so. I doubt whether I, as a young scientist isolated in Australia, would have had the courage to concentrate on that idea had it not been for the arrival of a paper from Lundegårdh (1945) in which he had reached the same idea though he had not analysed it quantitatively at that stage. I had been greatly influenced by Lundegårdh's perceptive approach to experiments on salt absorption and respiration and by his electro-chemical theory of salt absorption and respiration published in 1939. Lundegårdh, not accorded sufficient recognition in his lifetime, was certainly one of the greatest plant physiologists of this century. Almost three years after I had recognised a quantitative relation and formulated an hypothesis about the salt uptake mechanism, we were sufficiently satisfied with our quantitative data to publish (Robertson & Wilkins, 1948). The experimental material was consistent with the hypothesis that one anion might enter a cell as one electron went to oxygen, and one cation might enter as one proton was extruded. The same year we suggested, as did Conway & Brady (1948) and

106

Crane & Davies (1948*a*, *b*), that the hydrochloric acid secreted by the gastric mucosa might have a similar explanation. Both systems are much more complicated but, at the time, we were not to know.

When we advanced this hypothesis we did not know where cytochromes occurred in cells except that they were probably associated with particulate material. But, in 1948, Hogeboom, Schneider and Palade perfected the technique for the fractionation of cell organelles. Thanks to the work of Lehninger, Green and others, it was quickly shown that the reactions of respiration, from the stage where the original sugar molecule has been split and partially oxidised to two pyruvic acid molecules, all, including electron transport via cytochrome, were taking place in the mitochondria.

Our work moved rapidly into investigations of the ion movements in mitochondria for there, we thought, lay the explanation of our respiration-linked ion transport system. We certainly found that mitochondria can maintain ion contents differing from those outside and added our evidence for the presence of a membrane (Farrant, Robertson & Wilkins, 1953) in 1953. In trying to estimate its thickness from shadowed, collapsed mitochondria, we did not know that there were two membranes, an inner and an outer, and that the inner membrane had many folds. Later in the same year, Palade (1953) and Sjöstrand (1953), working with the newly developed technique necessary to cut sections of biological material for electron microscopy, showed that the mitochondrial structure is basically a smooth outer membrane surrounding an inner membrane which has many invaginations known as *cristae* (Figs. 1.1, 1.2). This was true of our plant mitochondria (Farrant, Potter & Robertson, 1956).

Techniques developed by many workers for the separation of outer and inner membranes made it possible to separate them and to characterise the enzymes which occur in the matrix inside the inner membrane, in the inner membrane, in the space between the membranes, and in the outer membrane (Fig. 6.4).

Fig. 6.4. A diagrammatic section through a mitochondrion showing the outer membrane (OM) and the very much folded inner membrane (IM) which forms the cristae (C).

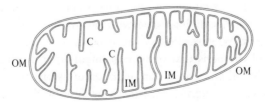

Inner membrane: charge separation

The intermediate compounds of respiration, starting with pyruvic acid, are successively relieved of their hydrogen atoms under the influence of the dehydrogenases which occur in the matrix of the mitochondrion. The series of reactions involved is the citric acid cycle but it is not necessary to describe that here. After the dehydrogenation, the two electrons from two hydrogen atoms are picked up by NAD^+, which also acquires one of

Fig. 6.5. The energy transduction of the inner mitochondrial membrane: (*a*) the first charge separation at FMN; (*b*) the ubiquinone acquires electrons and bobs up inside the membrane to pick up H^+; the uncharged ubiquinone then bobs down to lose electrons to cytochrome *b* and H^+ is lost to the outside; (*c*) the electrons pass through cytochrome b, c_1 to cytochrome *c* which passes them to cytochrome *a* and a_3; (*d*) cytochrome a_3 passes the electrons to oxygen and H_2O is formed; (*e*) by a mechanism unknown, two more protons pass from inside to outside the membrane; (*f*) protons pass through the ATPase to synthesise ATP inside.

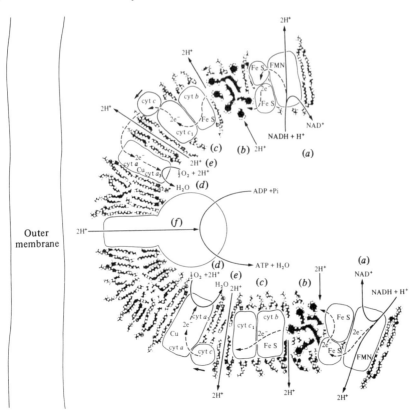

the two protons, and NADH is formed just as in the bacterial cell (Fig. 6.5). The other proton from the dehydrogenation appears to be loose in the water of the matrix. The NADH moves to the inner mitochondrial membrane to pass its two electrons to the membrane-bound carrier-group, flavine mononucleotide, FMN (which is analogous to the FAD of the bacterial cell). This reduced FMN then acquires the proton from the NADH and picks up one from the water. The electrons and protons pass across this membrane-attached protein associated with the FMN, the electrons going to an iron sulphur protein which is located near the external surface of the inner mitochondrial membrane. As the electrons depart from the FMN-protein complex, the protons are liberated to the *outside* of the membrane as hydrogen ions. Meanwhile, the two electrons are returned through the membrane towards the *inside*. There, they are passed to ubiquinone – either singly or in pairs – and the ubiquinone, on becoming charged, 'bobs up' in the membrane until its charged hydrophilic head picks up one proton for each electron, from the matrix of the mitochondrion (Robertson & Boardman, 1975). Uncharged ubiquinone then 'bobs down' again into the membrane until it comes into contact with cytochrome b which, though mostly buried in the membrane's lipophilic region, appears to have an electron accepting group at the lipophilic/hydrophilic interface. There, the two electrons leave the ubiquinone to pass into cytochrome b and the proton or protons are loosed in such a position as to make their way to the membrane's exterior. That means that the pair of electrons coming through the chain has already been associated with the transport of four protons to the outside of the membrane. The electrons continue on, passing through another cytochrome, c_1, and reach cytochrome c which, as we have seen (Chapter 2), is a peripheral protein, occurring in the watery space between the inner and outer membrane. From cytochrome c, the electrons pass to cytochrome oxidase, with its two

Fig. 6.6. The volume of the matrix (*a*) decreases and the space (*b*) between the inner and outer membranes increases during electron transport and phosphorylation.

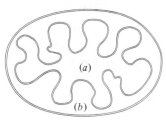

proteins cytochrome a and a_3 spanning the membrane. Cytochrome a_3 passes two electrons to oxygen in two successive reactions and, with two protons being acquired from the matrix solution, water is formed. By a mechanism which is still uncertain, two more protons pass from the *inside* to the *outside* of the mitochondrion somewhere between cytochrome b and the formation of water. I do not propose to confuse the principles by discussing the somewhat complicated investigations about this step. The experimental fact remains that the passage of two electrons through the electron-transport chain results in the extrusion of six protons which become hydrogen ions in the water between the inner and outer membranes. While this is going on there is usually a decrease in the volume of the matrix and an increase in the volume of the space between the membranes, particularly when the respiration rate is maximal in the presence of ADP (Fig. 6.6).

Inner membrane: oxidative phosphorylation

From analogy to the other membranes which have been discussed, the proton gradient which has been established by this respiration will be expected to be used for other processes. One of the most important is the synthesis of ATP from ADP and inorganic phosphate, and this occurs in the H^+-ATPase of the inner mitochondrial membrane. This ATPase has a hydrophilic knob (F_1) which is on the *inside* of the membrane and a lipophilic base (F_0) anchored in the bilayer (Fig. 6.7a). The protons on the *outside* of the membrane enter the F_0 base piece by an unknown mechanism and proceed by an unknown pathway to the F_1, where, provided that ADP and inorganic phosphate are present in the matrix, ATP is formed.

Since ATP is required for other cell processes, e.g. muscular contraction, ion secretion, etc. which occur outside the mitochondria, it is, at first sight, surprising that the synthesis occurs within the mitochondrion. The mechanism which compensates for this apparent difficulty is very interesting. Embedded in the inner membrane is a very specific carrier which is capable of picking up ADP on the outside and transferring it across the membrane to the *inside*, but only if it can transfer a molecule of ATP to the outside at the same time. This is a fascinating example of a *permease*, quite specific and membrane-bound, but not yet properly understood. Much of what is known about this enzyme has been deduced from the effects of its specific inhibitor, atractyloside, a glycoside which is somewhat similar to the ATP molecule. This is an example of a specific exchange diffusion in a membrane – one molecule in for one molecule out – and will be discussed further in Chapter 8. And so, ATP leaves the inner mitochondrial membrane as fast as it is formed. It can diffuse through the relatively permeable outer membrane to do its work in other parts of the cell.

Inner membrane: absorption and extrusion

The proton gradient which has been established has resulted in acid conditions outside the membrane and alkaline conditions inside. This difference is used by the mitochondrion to facilitate the entry and extrusion of the ions and molecules of metabolism. Just as ATP synthesis inside the mitochondrion requires ADP, it also requires inorganic phosphate, P_i. One of the substances (nature unknown) in the membrane can exchange a phosphate, $H_2PO_4^-$, for a hydroxyl ion from the interior. Thus the constant generation of OH^-, resulting from the electron transport chain, provides an exchange diffusion by this membrane-bound substance to keep up a

Fig. 6.7. The traffic in ions and phosphorylation in the inner mitochondrial membrane: (a) protons (extruded by the electron transport chain) enter the ATPase and synthesise the ATP from ADP which has entered on the nucleotide exchange carrier (b) and inorganic phosphate (P_i) which has entered in exchange for OH^- (c); P_i can exchange for succinate (d), succinate for malate (e) and malate for citrate (f); calcium enters via a carrier (g) and balances phosphate in the matrix; sodium leaves in exchange for protons (h); protons entering from the outside also function in the enzymatic reduction of $NADP^+$ by NADH (i) and in the entry of phosphate (j).

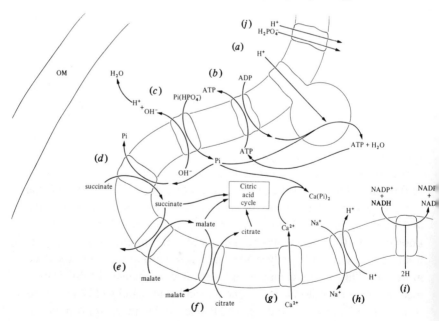

steady supply of $H_2PO_4^-$ for the phosphorylation mechanism. Various acids involved in the citric acid cycle within the mitochondrial matrix also pass through the membrane on carriers in exchange for another ion (Fig. 6.7). This will be discussed in Chapter 8.

The inner mitochondrial membrane, with its oxidation–reduction, charge separation, phosphorylation and ion movements, certainly must be one of the liveliest areas of cells, beautifully controlling the interacting processes. Fundamentally, these activities result from that basic property of energy-transducing membranes – the ability to play variations on the theme of charge separation and to turn the potential energy, so obtained, to various activities – the process of energy transduction. In the next two chapters we shall discuss how the different membranes, and different forms of energy, relate to ion and molecule movements.

The role of the lipids in energy-transducing membranes

In these two chapters we have discussed reactions which occur mainly in the proteins and particularly in pigmented proteins, but we have also seen how two highly lipophilic electron/proton carriers, plastoquinone and ubiquinone, function. Their activity depends not only on their solubility in the lipophilic region of the bilayer but also on their charged head groups which can make contact with the hydrophilic region to pick up protons. With the protons attached, the head groups become neutral, the quinones can move back into and diffuse rapidly through the lipophilic region. We have mentioned, a number of times, that the fluidity of the lipid hydrocarbons is essential for this and for other functions which include the capacity for a whole protein to alter its position (e.g. rhodopsin) and probably, though this is not shown experimentally, for a portion of a protein to move as a result of a conformational change. Further, the amphipathic properties of the membrane lipids provide the opportunity for a functional molecule, or group on a molecule, to 'see' either a lipophilic region or a hydrophilic region. For example, the two quinones, when in the quininoid, $HO-\langle\ \rangle-OH$, form, are very stable in a non-aqueous, non-polar environment. However, as soon as this group 'sees' a negative charge or water, it will become ionised and pass to the $^-O-\langle\ \rangle-O^-$ form. Similar changes in charge on proteins may be very important in moving a molecule vertically in the bilayer if it acquires a charge or tilting a molecule so a charged group on its side is brought nearer to the water.

One important question is why the different membranes contain such a wide range of different lipids, a matter referred to in Chapter 2. If the lipids functioned only to hold the membranes and their enzyme proteins together, it might be thought that one or two lipids might be enough to provide the necessary structure, packing, fluidity and so on. This is obviously not so. Not only are different lengths and different degrees of unsaturation in the hydrocarbon chains necessary to the membranes, but the conclusion that the head groups are necessary for particular functions is inescapable. It is presumably no accident that the inner and outer membranes of rat liver mitochondria have different compositions, for they have quite different functions. For example, both cardiolipin and phosphatidylcholine in the inner membrane are higher than in the outer. The lipid composition of the light-energy transducing membrane of the thylakoid is strikingly different from that of the oxidation–reduction transducing membrane of the mitochondrion; in thylakoids 80% of the lipid is galacto- or digalactolipid with little phospholipid, whereas in the inner mitochondrial membrane phosphatidylcholine and phosphatidyl-ethanolamine make up about 76% of the lipids.

We have discussed how such differences will affect the packing in the membrane and the fluidity of the different parts of the bilayer and, thereby, possibly have an important function in controlling the distribution of small lipophilic molecules within the bilayer. However, we are almost completely ignorant of the interactions which might take place between different head groups on the lipids in the immediate vicinity of an active group in a membrane-bound protein. If such interactions occur as part of the function of the lipid–protein enzyme complexes, their nature will not be understood until we know more about the protein structures and their conformational changes.

Suggested reading

Finean, J. B., Coleman, R. & Michell, R. H. (1978). *Membranes and their Cellular Functions*, 2nd edn. Oxford: Blackwell Scientific Publications (Chapter 4).

Harold, F. M. (1977). Membranes and energy transduction in bacteria. *Current Topics in Bioenergetics*, **6**, 83–149.

Hinkle, P. C. & McCarty, R. E. (1978). How cells make ATP. *Scientific American*, **238**, No. 3, March, 104–23.

7

Trans-membrane diffusion – ion carriers

In the first chapter, I introduced the idea that membranes are essential to the separation not only of the inside and outside of cells, but also to the compartments within cells such as the organelles. Membranes constitute an essential control mechanism in the cell's reactions just as surely as fences separate goats from crops or bulls from cows on a well-managed farm. They are the boundaries which make control of cellular operations possible, especially as so many cell reactions are between water-soluble molecules in water. As we have seen in the previous two chapters, they have other functions because the membrane-bound enzymes bring about reactions to leave one substance on one side of a membrane and another substance on the other side, reactions which would be impossible without the membrane as a barrier to diffusion. We have also seen that the hydrocarbon portion of the bilayer (Chapter 4) exercises control over what can enter, and hence what can pass through, and that the property of being very soluble in water is more or less incompatible with the ability to pass easily through the hydrocarbon portion of the bilayer. If the bilayer is continuous, the substance crossing must go out of solution in water on one side, dissolve in the membrane hydrocarbon layer and go into water solution on the other. If the bilayer is not continuous at the time a substance is passing, e.g. if it has water-filled pores, the diffusing substance may stay in water but its rate of movement can be restricted by the size and frequency of the pores. It is like the animal movements on a farm, controlled by a series of gates specially suited to the different sized animals and certainly opening and shutting when required, under the control of the farmer.

In this chapter we shall consider some of the complications which make membranes the barriers to diffusion and able to control the permeability of different types of substances. Some of these factors are the thickness of

113

114

the membrane, the mobility of its lipids and proteins, the solubilities of the permeating substance, both in water and in the membrane, and the length, breadth and tortuosity of any water-filled pores. Where ions are involved we must include the electric charges on the membrane surfaces, their signs and their densities. Last but not least, the total passage of a substance from one side to the other will be affected by the unstirred layer of water on each side of the membrane, stretching from that which is 'structured' water, held between the polar heads of the lipids, through that with normal kinetic movement to that which may be actively stirred, for example, by blood circulation in animals or protoplasmic streaming in plants.

It will be helpful to consider the diffusion of three types of substances: water, uncharged molecules and ions. The wide range of permeability coefficients (i.e. P, the amount crossing unit area of membrane in unit time with unit concentration difference on two sides of the membrane) is illustrated in Fig. 7.1. The range is about ten thousand million.

Fig. 7.1. The wide range of permeability coefficients. (From *Biochemistry*, 2nd edn, by Lubert Stryer. W. H. Freeman and Company. Copyright © 1981.)

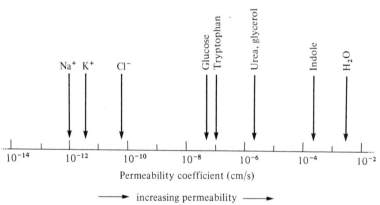

115

Water

In the latter half of the nineteenth century, plant physiologists, especially, were much preoccupied with unravelling their tangled thoughts about *osmosis* (i.e. the movement of water into plant cells without simultaneous loss of dissolved substances from the cell). About 100 years ago Pfeffer carried out the classical experiments with artificial 'membranes' which were rather different from our definition of membranes in living organisms for they were made by precipitating copper ferrocyanide in the spaces within a porous pot. But they served to establish the principles of 'semi-permeable' membranes because they were *much more* permeable to water than to dissolved substances. The extension of the principles by both plant and animal physiologists soon established that the protoplasm of a cell must either be semi-permeable or contain semi-permeable membranes to explain how water could diffuse in without, at the same time, appreciable loss of dissolved substances from the cell.

It was soon apparent that the osmotic entry of water into cells was not simply due to free permeability to water and complete impermeability to other substances for they, too, can enter though much more slowly. We have seen (Chapter 1 and Chapter 4) that the entry of water is likely to be due to the way in which water dissolves (albeit in very small amounts) in the lipophilic region of the membrane, so that some water molecules, few in number and isolated from each other, are in the centre of the bilayer. If the concentration of water on one side is increased (i.e. by diluting the solution on that side), water will pass through to the other side of the

Fig. 7.2. A diagrammatic representation of the restrictive effect of a monolayer of the C_{16}-alcohol, cetyl alcohol, on evaporation from a water surface: (*a*) with free water; (*b*) with a layer of cetyl alcohol.

(*a*)   (*b*)

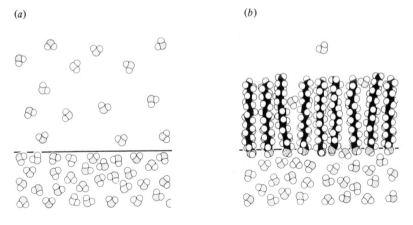

membrane until equilibrium of water concentration is attained in relation to any pressure difference existing on the two sides. If the hydrocarbon part of the membrane is interrupted by water-filled pores or by channels of predominantly hydrophilic groups, water will have no difficulty in passing through. Water passage into and out of the membrane will be only minimally affected by the unstirred layer and by the charges on the polar heads since one water molecule in these regions can be easily displaced by another.

Though the movement of water across bilayers is rapid compared with that of other molecules in aqueous solution and especially rapid compared with that of ions, the effects of the hydrocarbon layer in amphipathic molecules in restricting the diffusion can be quite striking. For instance, a monolayer of cetyl alcohol (C_{16}) is very effective in reducing the evaporation of water into the air (Fig. 7.2). Indeed, many experiments were carried out in dry Australia to try to use cetyl alcohol as a monolayer on dams or tanks to restrict water loss by evaporation (Mansfield, 1958). Its effectiveness in restricting evaporation was undoubted but its practical usefulness was ruled out by the difficulty of maintaining the layer, since the cetyl alcohol was blown to the sides and precipitated on the banks or walls.

Water can be expected to pass through membranes quite rapidly compared to most other molecules but the lipid–protein bilayer slows down its diffusion rate.

Membrane diffusion of uncharged molecules

The pioneer in recognising that fat or oil solubility was important in allowing molecules to enter cells was Overton (1895). Having observed that molecules with some fat solubility enter cells from solution in water, he concluded that the limiting layer of the protoplasm must be impregnated with a substance with powers of solution similar to fats – a far-sighted hypothesis considering the ignorance of the times, but not without rival hypotheses. Küster (1911), and especially Ruhland (1912), had observed that smaller molecules of dyes entered cells more easily than larger molecules and suggested that the control of entry was due to ultra-filtration. Some years later Collander & Bärlund (1933) obtained evidence that smaller molecules entered more rapidly than would be suggested by their lipid solubility alone and they combined features of both hypotheses as the lipid-sieve hypothesis. In retrospect, with our understanding of bilayers today, we can see that the ideas of both Overton and Ruhland were right in part and that the differences are to be expected when we consider the

117

hydrophobic portions of the bilayers, and the movements of the lipid molecules depending on their fluidity.

The idea that a bilayer of molecules, which themselves share a degree of mobility, can be an effective barrier to the diffusion of small molecules is sometimes surprising to people hearing about it for the first time. I have found that those attending my lectures have been greatly helped to visualise the resistance of a membrane by a simple experiment which originated with Dr Bangham of the Institute of Animal Physiology (Babraham, UK), an experiment we have called the 'Bangham flambé experiment'. If two trays are set up, each with a saturated solution of diethyl ether in water, the volatile ether evaporates from the water surface to the air. (*These experiments should not be done by those inexperienced in handling ether with the usual precautions.*) A match applied to the surface of each tray results in a merry blaze, rather better than that of brandy on the Christmas pudding (Fig. 7.3). If, now, a drop of soap solution, with its surface-active molecules of sodium palmitate, is applied to one of the trays, the soap molecules spread rapidly across the air–water interface as a monolayer and

Fig. 7.3. The restrictive effect of a layer of soap molecules on the diffusion of diethyl ether: (*a*) a saturated solution of ether in water in each tray has a match applied; (*b*) the ether diffusing from each tray burns effectively; (*c*) a drop of soap solution is applied to the left tray and the burning is cut off, whereas the control (right tray) continues to burn; (*d*) the left tray will not re-ignite, the right tray continues to burn until the ether is exhausted.

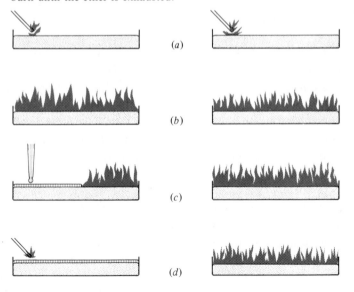

immediately extinguish the flame. Meanwhile the other tray continues to blaze until the ether has been used up. This striking effectiveness of a single layer of amphipathic molecules at the water–air interface in stopping the diffusion of the small ether molecules demonstrates how efficient similar molecules packed into a bilayer can be in resisting diffusion and therefore acting as a barrier in a cell.

Passive diffusion

The permeability of molecules which we have been discussing is known as *passive permeability*, meaning that it results from a concentration gradient and is due only to *passive diffusion* resulting from the kinetic movements of the diffusing molecules and the membrane molecules. Not surprisingly, it is very dependent on the lipid composition of the bilayer since, as we have seen, this governs the molecular movement of the lipid hydrocarbon chains. For instance, the presence of cholesterol in an artificial membrane (De Gier, Mandersloot & Van Deenen, 1968) or in natural membranes (Grunze & Deuticke, 1974) has been shown to decrease the solute permeability. Similarly a high degree of unsaturation in the hydrocarbon chains will increase permeability (De Gier *et al.*, 1968). Thus, as we have discussed in Chapter 4, the forces holding the membrane molecules together will exert considerable control on the rate of penetration of diffusing molecules. Furthermore, the difference can exist within different parts of a bilayer. In erythrocytes this difference is seen between the two monolayers which, together, form the bilayer; the outer contains cholesterol, and choline phospholipids; the inner does not have much cholesterol but contains phosphatidylethanolamine and phosphatidyl-serine (Verkleij *et al.*, 1973). A difference will also occur if some portions of the hydrocarbon chain, such as the first eight carbons adjacent to the polar groups, are held tightly by van der Waals forces though the centre portions of the same molcules, due to unsaturation, are still fluid.

Permeability is also affected by temperature; in general, the higher the temperature the greater the diffusion rate of solute, the greater the molecular movement in the bilayer and the greater the permeability. As the temperature is lowered to the phase-transition temperature, i.e. the temperature at which some of the bilayer molecules pass to the solid state, the permeability may, paradoxically, be increased (Haest *et al.*, 1972; Blok *et al.*, 1975). This permeability increase appears to be related to the co-existence of solid and fluid phases in the bilayer. As one lipid constituent begins to solidify, the mobility of another lipid may be increased because of changes in packing.

In Chapter 3, I discussed how the presence of proteins in the bilayer can have a perturbing effect on the lipids. If this perturbing effect results in greater fluidity in the lipid molecules, permeability increase will be expected. Various experiments have shown that liposomes have increased permeability when protein molecules are introduced (Haydon & Hladky, 1972). On the other hand, if a lipophilic protein holds an annulus of immobilised lipids which in turn immobilise adjacent lipids, the membrane molecule mobility may be decreased and the permeability lowered.

A substance in solution will cross a membrane if its concentration on one side is greater than the concentration on the other, following the simple principles of diffusion. The process proceeds with a change in free energy until equilibrium is reached with equality of concentration on each side. This change in free energy is proportional to the log of the concentration ratio, i.e.

$$-\Delta G = RT \log C_r/C_l$$

where $-\Delta G$ is the change in free energy, R is the gas constant, T the

Fig. 7.4. Passive diffusion and ion exchange: (a) glucose in solution in water diffuses through a permeable membrane (left) until the concentration is equal on both sides (right); (b) K^+ and Cl^- both diffuse from the side of high concentration to equal concentration on each side; (c) exchange of cations will occur until the electrochemical potentials on each side are equal.

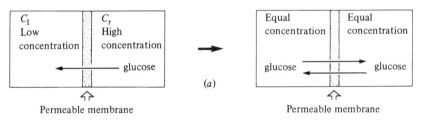

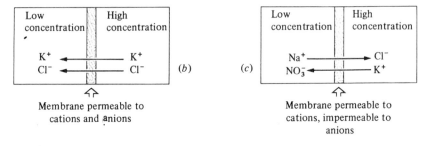

absolute temperature, C_r the higher concentration and C_l the lower concentration (Fig. 7.4a). As long as such diffusion is proceeding towards equilibrium there is a decrease in free energy.

Equilibrium does not mean that none of the molecules are passing across the membrane; they are, but the number crossing the membrane from the left becomes equal to the number crossing the membrane from the right and no net movement from one side to another is detected unless a special experiment is set up. If the molecules of the substance on the left, e.g. glucose, are replaced by glucose molecules labelled with the radioactive isotope ^{14}C, the exchange of labelled molecules for non-labelled molecules across the membrane can be observed by the increase in radioactivity on the right and the decrease on the left until both sides have equal radioactivity. The movement of molecules across a membrane is known as the *flux* and is the amount crossing in unit time. It can be obtained by multiplying the permeability, P, by the concentration differences between the two sides. Thus the flux, ϕ and C_l in Fig. 7.4a becomes

$$\phi_{r\,to\,l} = P(C_r - C_l)$$

at any stage of the diffusion. It has the same units as P, i.e. mol cm^{-2} s^{-1}. We can speak of the *influx* of a molecule into and the *efflux* of a molecule out of a cell or organelle. As both are going on at the same time, net influx (ϕ) would equal influx (ϕ_{oi}) minus efflux (ϕ_{io}).

Net flux would be

$$\phi = \underset{\text{influx}}{\phi_{oi}} - \underset{\text{efflux}}{\phi_{io}}$$

and would be positive when influx exceeds efflux, zero at equilibrium, and negative if efflux exceeds influx.

Facilitated diffusion

So far we have been thinking about the simple diffusion resulting from the entry of a particular molecule without the specific intervention of a membrane-bound molecule. These considerations apply particularly to small molecules but, when we consider molecules of the size of hexoses, we note that the passive permeability is quite small compared with that of water: compare the permeability coefficient of water in Fig. 7.1 with that of glucose whose permeability coefficient is about 100 000 times less. The biological advantage of very impermeable membranes is that they allow the cell, not only to control the distribution of molecules and ions, but also to handle them specifically. It is not surprising, therefore, to discover that not many of the molecules and ions in living systems pass through membranes by simple diffusion but do so via membrane-bound transport systems which are quite specific. This is called *facilitated diffusion* or *mediated permeation*. Extensive studies with *E. coli* have shown that specific *carriers* or *permeases*, which will transport a particular sugar across the membrane in a normal cell, may be absent in a mutant. Such permeases are known, for example, for glucose, maltose and lactose. They are membrane-bound proteins (that for lactose has a molecular weight of about 30 000) and their genetic controls on the bacterial chromosome have been mapped. Another carrier system of *E. coli* is responsible for glycerol entry and, in *Streptococcus faecalis*, there is a lactic-acid transport molecule. We have already referred (Chapter 6) to the mitochondrial permease which is responsible for taking an ATP out in exchange for an ADP in.

Facilitated diffusion takes different forms which are described in the widely-used terminology introduced by Mitchell. If a single substance is carried across the membrane by the permease, the system is known as a *uniport*, as with the sugars described above. If the substance carried in through the membrane is balanced by a specific substance carried out, i.e. facilitated exchange diffusion, it is an *antiport*. All the organic acid movements shown in Fig. 6.7 are antiports. When a substance is carried across accompanied by another substance it is termed a *symport*, e.g. H^+ and $H_2PO_4^-$ in Fig. 6.7.

Permeases are proteins which resemble enzyme molecules since they:

(1) are stereospecific for the molecule transported;

(2) are inhibited by structural analogues which can bind to the same site on the protein, and

(3) become saturated with the transported molecule at high concentrations, as in the relation between enzyme and substrate, resulting in a maximum rate. Just how the permeases operate in the membrane to bring a molecule from one side to the other is not known.

Three possible actions have been suggested:

(1) the carrier molecule, which is a large protein, picks up the transported molecule on one side of the membrane and rotates to let it go on the other side; this is very unlikely because of the constraints in 'flip-flop' (see Chapter 4), and the difficulty of breaking the hydrophilic bonds on two sides of the protein to allow it to rotate. Not all proteins are so constrained in other movements within the membrane; some do seem to be able to move up and down (e.g. rhodopsin in retinal membranes), or to rotate about

an axis perpendicular to the plane of the membrane (e.g. the Ca^{2+} ATPase of the sarcoplasmic reticulum), but not to rotate from one side to the other;

(2)   the carrier molecule undergoes a conformational change after it has picked up the transported molecule in water and then liberates it in a part of the membrane where it can go only to the other side;

(3)   the protein molecule, either alone or through cooperation with other membrane molecules, forms a channel through which the transported molecule can diffuse, either because it is a water-filled channel or because there is a possible succession of hydrogen bonds or charged residues.

Whatever the mechanism, facilitated diffusion for a particular molecule is usually much more rapid than if that molecule were moving by passive diffusion. For example, glucose crosses the membrane of the erythrocyte about 100 000 times more rapidly by facilitated diffusion than it could by passive diffusion. This is because the activation energy in combining with its carrier is much less than the energy required for the passage of a glucose molecule, with its strong hydrogen bonding to water, into the lipophilic membrane. The common phenomenon of carrier-mediated molecular transport is often dependent on a chemical reaction for provision of energy. We shall return to this aspect in talking about transport, but first we must discuss diffusion of ions.

Ions

I have already commented on the much lower permeability of membranes to ions than to small molecules and the very much lower permeability to ions than to water. Reference to Fig. 7.1 again reminds us that the diffusion coefficients of Na^+ and K^+ in membranes are about ten orders of magnitude (10 000 million times) less than that of water, and about five orders of magnitude (100 000 times) less than that of glucose. But, with ions, we cannot consider an ion of one charge without considering the ion of the opposite charge when we are talking of diffusion from water on one side of a membrane to water on the other. Thus, if K^+ is to leave a more concentrated solution of KCl on one side of a membrane to diffuse to the other side, it must either be accompanied by a Cl^- or it must replace a positive ion already on the other side balanced by another kind of anion, e.g. nitrate (Fig. 7.4b, c).

Charges on ions result in electrical potential differences in circumstances where the positive and negative charges are not exactly balanced or are not moving at the same rate. Indeed, the mobilities of positive and negative ions are seldom equal even in water and, in consequence, *a diffusion potential* is established. For example, if the salt KCl is diffusing in water from higher to lower concentration, a potential difference can be measured because the mobility of the K^+ ion in water is greater than that of the Cl^-. At all liquid junctions a diffusion potential is set up since one ion of a salt invariably tends to cross more rapidly than the other ion. Referring to Fig. 7.4*b*, imagine that the mobility of K^+ in the membrane is greater than that of Cl^-, the K^+ 'tending to get ahead of' Cl^- would make the left side more positive. If the membrane is permeable to K^+ and negligibly permeable to Cl^-, a potential difference of 55 mV is established at 15 °C with a tenfold concentration drop between the right and the left. The relationship is given by

$$E_r - E_l = \frac{\mu^+ - \mu^-}{\mu^+ + \mu^-} \frac{RT}{F} \ln \frac{C_l}{C_r}$$

where E_r and E_l are electrical potentials relative to an arbitrary zero, μ^+ and μ^- are the mobilities of cation and anion respectively in the electric field, R is the gas constant, T the absolute temperature, F the Faraday and C_l and C_r the concentrations on left and right respectively. Concentrations, especially at higher values, need correcting for activities.

We have seen that the decrease in free energy of an uncharged molecule diffusing from high to lower concentration depends on the concentration ratio, but the decrease in free energy of charged species must take account of the *electrochemical potential difference* across the membrane. This potential includes two driving forces acting on the ion:

(1)   the gradient of the chemical activity (often approximately the same as gradient of concentration), and

(2)   the electric potential difference.

Only when both are taken into account in diffusion of charged particles can we see that ions move towards an equilibrium, not of equal concentrations on each side of a membrane, but of equal electrochemical potentials. In this case the change in free energy is given by

$$-\Delta G = RT \ln C_r/C_l + ZF \, \Delta V$$

where Z is the electrical charge on the ion diffusing, F is the Faraday and ΔV is the potential across the membrane. In the simple example illustrated in Fig. 7.4*b*, K^+ and Na^+ can exchange across the membrane but will do so only until the electrochemical potentials on each side are equal and, at that point, no current will flow. The total current density will be the sum of the currents due to different ions and is zero at equilibrium.

It can be shown (see Briggs, Hope & Robertson, 1961 or Hope, 1971, for the full development), that the potential difference is related to the permeabilities and concentrations of the different ions. Referring to the example in Fig. 7.4c, the potential difference across the membrane would be

$$E_r - E_1 = RT \ln \left(\frac{P_K K_1}{P_K K_r} + \frac{P_{Na} Na_1}{P_{Na} Na_r} + \frac{P_{Cl} Cl_1}{P_{Cl} Cl_r} + \frac{P_{NO_3} NO_{3_1}}{P_{NO_3} NO_{3_r}} \right)$$

where K_1, Na_1, Cl_1 and NO_{3_1} represent concentrations on the left side, K_r, Na_r, Cl_r and NO_{3_r} concentrations on the right side, P_K, P_{Na}, P_{Cl} and P_{NO_3} are the respective permeabilities and R, T and F have their usual meanings.

This discussion serves to highlight the two important differences between charged and uncharged substances crossing a membrane:

(1) charged particles do not necessarily reach equilibrium at equality of concentration;

(2) ions move independently of each other, depending on the electro-chemical potential gradient.

As we shall see in the next chapter, moving an ion against its electrochemical potential gradient requires energy and work must be done. In natural systems, the electrochemical potential gradient may be due to the total effects of many different ions.

Donnan equilibrium

In living systems, and membranes natural or artificial, we have to consider ions which are restrained or indiffusible, i.e. those which are attached to some part of the solid structure of an organelle or a membrane and cannot diffuse. The same effect is seen if a membrane is almost completely impermeable to ions of one charge, e.g. in Fig. 7.4c, a system which would apply if the membrane were completely impermeable to Cl^- and NO_3^-. The effects of indiffusible ions on diffusible ions in their neighbourhood was first considered by Donnan in 1911 although implicit in Gibbs' equations derived in 1906 for heterogeneous equilibria. The state of all the ions at equilibrium is known as the *Donnan equilibrium*.

Effects associated with a Donnan equilibrium can be measured and we find:

(1) that the indiffusible ions attract the diffusible ions of opposite charge;

(2) that an electric potential difference exists between the region of the indiffusible ions and the adjoining solution where the ions are diffusible;

(3) changing the kind of diffusible ions results in ion exchange with those already balancing the indiffusible ions.

In membranes themselves, we have seen that there are indiffusible ions, particularly those of the phospholipids, fixed in the membrane. Most membranes have a predominance of fixed or indiffusible negative charges so that positive charges are attracted to, and negative charges repelled from, the membrane surfaces. This layer of negative indiffusible ions and the adjoining crowd of positively charged mobile ions form what is called the *electric double layer*. Close to the surface of negative charges there will be more diffusible cations than diffusible anions, but the difference will decrease with distance from the membrane surface. The electric double layer has a profound effect on the ions in the vicinity and is itself modified substantially by the charge of the diffusible ions (monovalent, divalent, etc.) and by their concentration. It governs the ions that can and do approach the charged parts of the membrane surface.

Weak electrolytes

The permeability of weak electrolytes, which may be in aqueous solution in unionised form, is different from that of ions and is absolutely dependent on pH. Weak acids, e.g. acetic, lactic, succinic, have little inclination to dissociate in water, i.e. to donate their protons to a water molecule with the formation of a hydroxonium ion. This means that, under most conditions, molecules of undissociated acid are present together with positively charged hydroxoniums and negatively charged anions of the acid. This contrasts with strong acids in which the dissociation is more nearly complete. In the reaction

$$HA \rightleftharpoons H^+ + A^-$$

where HA is the undissociated acid, H^+ the hydrogen ion and A^- the anion, the apparent equilibrium constant will be

$$K = \frac{[H^+][A^-]}{[HA]}$$

where the square brackets represent concentrations. The lower the pH (i.e. the more H^+), the more the reaction will be driven to the left. Conversely, the higher the pH, the less H^+, the more the reaction will be driven to the right.

A useful constant for a weak electrocyte is the pK, defined as $-\log K$ or $\log 1/K$, and corresponding to the pH at which the acid is 50% dissociated. The pK is very valuable in deciding what ions or molecules will be in the solution of weak electrolytes. For example, the pK of succinic acid is 4.2, i.e. at a pH of 4.2 succinic acid is 50% molecules and 50% succinate and hydrogen ions. If the pH is lowered, the proportion of ions

126

decreases; if it is raised the proportion increases. The pK of H_3PO_4 is 2.14 which means that the pH must be below that to achieve more H_3PO_4 than $H_2PO_4^-$ and H^+.

Just as acids are proton donors and dissociate in water, bases (B), e.g. ammonia, are proton acceptors, thus

$$BH^+ \rightleftharpoons B + H^+$$

e.g.

$$NH_4^+ \rightleftharpoons NH_3 + H^+.$$

Bases also have a dissociation constant and a pK which tells us at what pH the base is 50% protonated. The pK of ammonium is 9.25, i.e. only above pH 9.25 is there less than half the ammonia protonated. The pK of methylamine is 10.6.

Amphoteric electrolytes, e.g. amino acids and some lipids, which have both an acid group and a base group, have two pKs, one for the acid and one for the base. For instance, the pK of glycine is 2.3 for the —COOH group and 9.9 for the —NH$_2$ group. This means that in the pH range 2.3–9.9, the acid group will be mostly —COO$^-$ and the amino group will be mostly —NH$_3^+$, the two charges compensating each other as a zwitterion and the whole molecule neutral.

We have seen that ions do not enter membranes as readily as neutral molecules. If the ionisation can be suppressed, more of a substance can penetrate the membrane (Fig. 7.5). When an undissociated acid penetrates and ionises on the other side at a higher pH, the escape of ions through the membrane will be prevented. There are innumerable examples in which the entry of a weak acid into a cell, e.g. an acid of the Krebs cycle such as succinate, can be accelerated by lowering the external pH. Though it

Fig. 7.5. Weak electrolytes cross living membranes as undissociated molecules: (a) diffusion of undissociated acid (HA) from region of low pH to region of high pH results in (b) an increase of anion (A$^-$) balancing the cation (K$^+$).

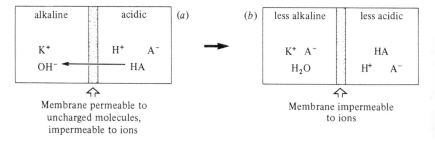

is the molecule that penetrates the membrane, it is often the ion which is effective in the reaction inside. In the next chapter we shall see an example of the entry of an unprotonated, uncharged base into chromaffin granules where it becomes protonated and cannot go back through the membrane because of its charge.

Ion carriers and channels

Since the barrier to movement of ions, which allows the cell or organelle to control its composition, is largely the lipophilic hydrocarbon layer, it follows that an ion disguised in a lipophilic parcel could be smuggled through. This principle has been known for a long time because many small molecules with lipid solubility, 2,4-dinitrophenol, CCCP etc., carry protons through membranes. If they do so under circumstances in which a proton gradient is normal and cause its collapse, they are said to be uncouplers. They were first used to prevent oxidative phosphorylation

Fig. 7.6. The ionophore which is the transport antibiotic, valinomycin in a bilayer. Note that the ring of four acids, each repeated three times, results in a lipophilic doughnut which can trap K^+ in its centre.

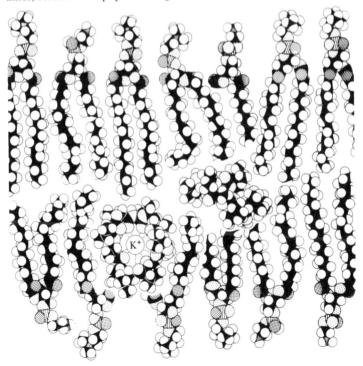

in mitochondrial experiments. Such molecules, which carry a proton across a membrane, are termed *protonophores*; those which carry other ions are known as *ionophores*.

Curiously enough this principle has been developed in evolution as a defence mechanism for some microorganisms which produce molecules that make membranes of other organisms permeable to certain ions. These *transport antibiotics*, as they are called, may be either carriers which dissolve in the membrane because of their lipophilic properties or channels which allow an ion to pass right through. These molecules, not harmful to the organisms producing them, kill other organisms by making their membranes permeable so their energy relations are upset and the ATP synthesis (Chapters 5, 6) is prevented. For example, one such antibiotic, valinomycin, makes the membrane permeable to K^+ and the charge separation which would make ATP is wasted on accumulating K^+. Another, nigericin, carries K^+ one way and H^+ the other so the proton gradient collapses.

These membrane-puncturing antibiotics have organic acid residues arranged so that their lipophilic groups are on the outside of a ring like a doughnut and several oxygen atoms are towards the centre of the ring. Valinomycin (see Fig. 4.8), for instance, consists of four acids, L-lactate, L-valine, D-hydroxyisovalerate and D-valine. These four residues are repeated three times round the ring with their hydrocarbon groups towards the outside. K^+ loses its water of hydration and becomes co-ordinated to six oxygen atoms in the centre of the ring and is, in effect, wrapped up in lipophilicity and the valinomycin complex can enter the membrane (Fig. 7.6). On passing into the aqueous zone on the other side of the lipid region, competition for K^+ between water molecules and the valinomycin results in hydrated K^+ leaving the valinomycin to go into the water. Valinomycin in concentration of only 10^{-7} M can lower the resistance of a bilayer to K^+ by a factor of 10000 with a gradient of 0.02 M KCl across the membrane.

The channel-forming antibiotics act differently. The best known is gramicidin A which is an open chain polypeptide of 15 amino-acid residues arranged in a cylinder with a channel about 0.2 nm in diameter through its long axis. The hydrophobic side chains of the amino acids are on the outside of the cylinder. Two cylinders, end to end, will span a membrane and render it very permeable to cations but not to anions (Fig. 7.7). The cations move through the water-filled channel, not particularly specific to the cations, but very efficient. It has been shown that transport rates of up to 2×10^7 cations per second can occur.

It is significant that the molecules which have been designed in the course of evolution to make membrane penetration of ions fast and easy should be weapons of war. The implication is that cells do not, as a rule, want ions to cross their membranes easily. All the cell and organelle control systems depend on the effective barrier to those ions which are not required to enter or leave through the membrane at a particular time. The living organism must control specifically what ionic movements are required at any one time. How this is done will be the subject of the next chapter.

Suggested reading

Finean, J. B., Coleman, R. & Michell, R. H. (1978). *Membranes and their Cellular Functions*, 2nd edn. Oxford: Blackwell Scientific Publications (Chapters 3, 4).

Quinn, P. J. (1976). *The Molecular Biology of Cell Membranes*. London and Basingstoke: The Macmillan Press Ltd. (Chapter 4).

Fig. 7.7. The transport antibiotic, gramicidin, which is lipophilic on its outside so it sits in the membrane, has a central channel through which cations can pass readily. The diagram shows that each dimer has 15 amino acids (1, valine; 2, glycine; 3, alanine; 4, leucine; 5, tryptophan); hydrophilic side chains are in the centre (hatched).

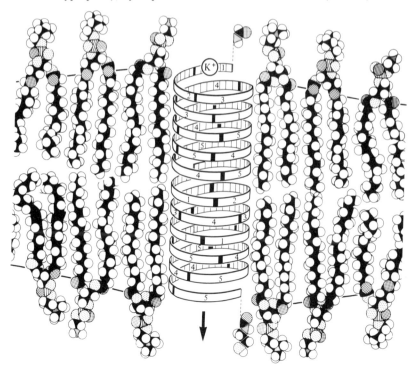

8

Transport, absorption and secretion

We have seen that simple diffusion is unlikely to be as advantageous to the cell's control of its environment as would be specific transport processes. It is not surprising then, to find a variety of transport mechanisms suited to the different systems essential to cells.

We have already spoken about several specific transport systems – the particular relation of phosphatidylcholine in membranes to the movement of H^+ and Cl^- (Chapter 4), the mechanisms of proton translocation (Chapters 5 and 6), and the carriers which transport molecules (Chapter 7). In this chapter we shall consider some examples of transport systems and relate them to processes of absorption into and secretion from cells.

Group translocations

The uptake of sugars, particularly well understood in bacterial cells, is of two kinds. The first has been discussed already and related to the proton gradient. The second is accompanied by a change in the molecule when it is transferred across the membrane (Roseman, 1972). Glucose, for instance, is converted into glucose-6-phosphate as it crosses the membrane in *E. coli*, *Salmonella typhimurium*, and *Staphylococcus*

Fig. 8.1. A possible schema for sugar transport by the bacterial cytoplasmic membrane: (*a*) glucose (G) outside the membrane, enzyme I (E I), phosphoenolpyruvate (*P*EP) and carrier protein (HPr) inside; (*b*) glucose moves in and combines with enzyme IIA (E IIA) in the membrane, *P*EP and HPr form a complex with E I and phosphate (*P*) is transferred to HPr from pyruvate (*P*y); (*c*) glucose on E IIA in the membrane and phosphorylated HPr (*P*-HPr) combine with E IIB; (*d*) complex of E IIA, E IIB, G and *P*-HPr is formed; (*e*) glucose-6-phosphate (G*P*) and HPr leave membrane to the inside; (*f*) G*P* is trapped inside cell.

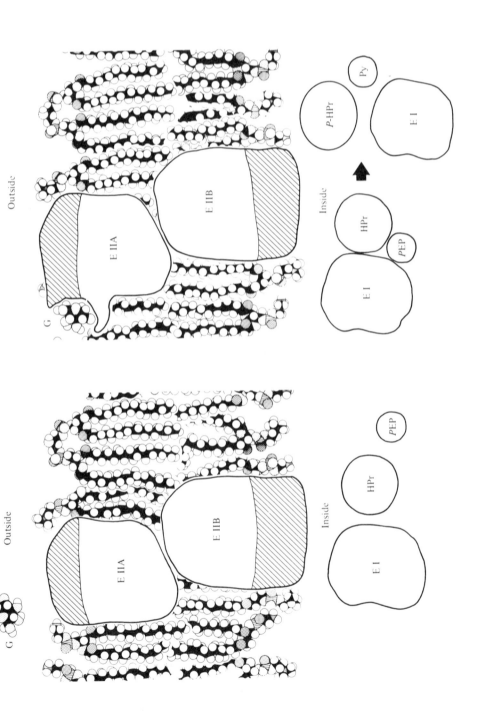

132

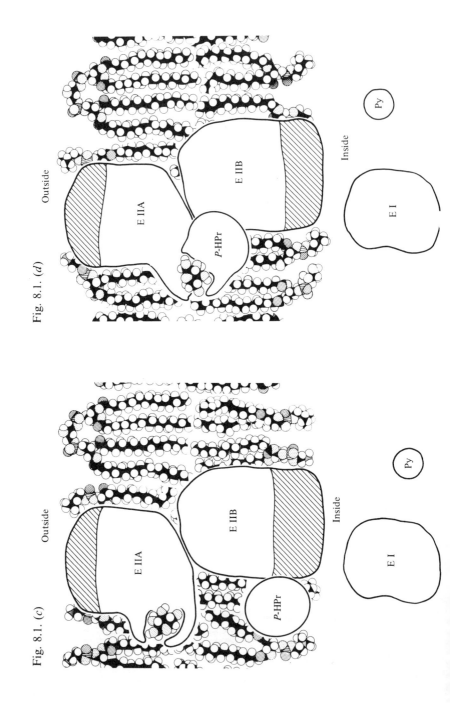

Fig. 8.1. (c)

Fig. 8.1. (d)

133

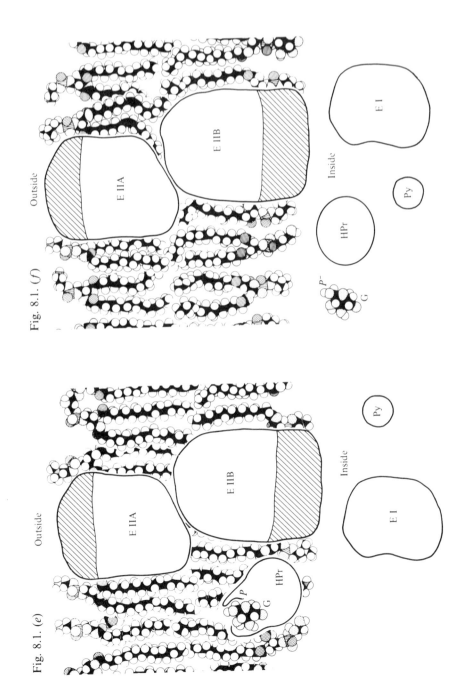

Fig. 8.1. (e)

Fig. 8.1. (f)

aureus. This type of uptake is known as *group translocation* and is brought about by a membrane-bound enzyme complex called phosphotransferase. The phosphate for phosphorylating the glucose is donated by phosphoenolpyruvate supplied by the cell and the directional (or vectorial or sided) reaction results in glucose picked up outside the membrane being phosphorylated from the phosphoenolpyruvate inside the membrane (Fig. 8.1). Four proteins are involved in this system. The energy for the reaction comes from the dephosphorylation of phosphoenolpyruvate, catalysed by enzyme I, and the phosphate is transferred to a particular histidine of a small carrier protein referred to as HPr with molecular weight of 9600 (Fig. 8.1*a,b*). This phosphorylated carrier protein then forms a complex with enzyme II which is built into the membrane (Fig. 8.1*c, d*) and, with the incoming glucose molecule, results in the transfer of the phosphate to form glucose-6-phosphate (Fig. 8.1*e*). This molecule is then liberated to the interior of the cell from which it cannot escape because its charge renders it unable to penetrate the cell membrane. The vectorial arrangement is clear; glucose-6-phosphate when formed is expelled to the inside of the membrane but not to the outside (Fig. 8.1*f*).

Enzyme II is particularly interesting because it consists of two functional parts, A and B. Enzyme II A, in occurrences of the enzyme in different organisms, is specific for a particular sugar, i.e. there is a II A for glucose, another specific for fructose and another specific for lactose. In mutants without the specific enzyme II A for a particular sugar, that sugar is not transported. Enzyme II B, the carrier protein and enzyme I are common to all the sugar reactions and, if they are absent in a mutant, no sugars are transported.

This system transporting the sugar group provides an example of interaction between intrinsic and extrinsic proteins. Enzyme I and the carrier protein are soluble in water but enzyme II is membrane-bound and the complex requires lipids for its activity. Carrier protein must form a complex with enzyme II A and B which are in the membrane, and with the sugar which enters. When the sugar phosphate is formed it can go only in, not back to the exterior. What particular membrane structures are associated with this control are not known.

Active transport

Systems which can transport molecules against their concentration gradients or ions against their electrochemical potential gradients are called *active transport* systems. As we shall see, active transport in one form or more is to be found in every living cell. Such systems use energy provided by the cell to work against the tendency for everything to reach chemical equilibrium. In chemical systems, there is a steady tendency to decrease the free energy and increase the entropy. Everything tends to equilibrium as surely as water runs downhill; if, however, water running downhill is used to turn a wheel, energy can be transferred, as in the old water mills, for turning the stones which ground the flour. Living cells use the energy running down through respiration to provide that which can be stored and used in other processes which require it. In this section we shall see how energy storage or transfer systems, can bring about the energy-requiring processes of active transport.

Formally, active transport is a reversal of the decrease in free energy which occurs when concentration or electrochemical systems tend toward equilibrium. We have seen in the last chapter that the change is summarised in the equations

$$-\Delta G = RT \ln \frac{C_r}{C_1} \text{ for uncharged molecules}$$

and

$$-\Delta G = RT \ln \frac{C_r}{C_1} + ZF \, \Delta V \text{ for ions.}$$

In spontaneous, non-energy requiring processes, ΔG is negative, i.e. there is a decrease in free energy. In active transport, which is proceeding *away from* the equilibrium condition, the free energy in the system must be increased, i.e. ΔG must be positive and energy must be supplied.

Active transport systems take different forms and must be looked at individually; some are dependent on the proton gradient, some on other ionic gradients, some on ATP operating specific ATPases.

Active transport processes dependent on charge separation

The very great importance of charge separation and the consequent proton gradients in active transport was greatly illumined by the imaginative work of Peter Mitchell with his chemiosmotic theory from 1961 onwards. This theory is fundamentally about the transport of ions and molecules in cells and tissues. It is interesting to observe the history of opposition to Mitchell's ideas, largely because the biochemists of the day, used to thinking of metabolism by soluble enzymes and metabolic intermediates, found it difficult to accept not only that so much happened in membranes, but also that the membrane reactions had vectorial pathways or sidedness. Furthermore, the research of the day was preoccupied with finding a phosphorylation system, at the electron transport chain, which resembled that of non-oxidative phosphorylation. There was also a widespread idea that the formation of ATP must precede and cause the charge separation (see Robertson, 1960).

The essentials of the Mitchell hypothesis relating to ion gradients are:

(1) the membrane with low permeability to ions;

(2) the charge separation which results in protons going to one side and electrons to the other and is followed by hydrogen ions in the water on one side and hydroxyl ions on the other;

(3) specific exchange diffusion systems which allow the positive and negative charges (hydrogens or hydroxyls) to be swapped for cations and anions respectively;

(4) a membrane-bound ATPase which, as a result of an H^+ gradient, can synthesise ATP from ADP and P_i but which, conversely, can hydrolyse ATP and establish an H^+ gradient.

Mitochondrial transport

The Mitchell hypothesis was largely developed in relation to the proton gradient and potential difference occurring across the inner mitochondrial membrane, and immediately helped to explain a number of the many experimental observations on mitochondrial ion systems. It will be remembered (Chapter 6) that protons are extruded from a respiring mitochondrion; simultaneously there is an accumulation of hydroxyl ions in the matrix inside.

To understand the way in which a proton gradient brings about redistribution of cations and anions, it helps to consider a very simple system (Fig. 8.2).

When hydrogen ions have been extruded to the outside and hydroxyls accumulated to the inside (Fig. 8.2a), two things can be expected:

(1) if the membrane is permeable to cations, external cations will move to the interior to balance the hydroxyls and maintain electrostatic neutrality;

(2) if the membrane is permeable to anions, the OH^- from the interior will move out in exchange for anions from the exterior.

If the concentration of OH^- on the inside of a cation-permeable membrane exceeded the concentration of the salt in the external solution, the cations would enter spontaneously without further energy requirement (Fig. 8.2b). If, now, the uptake of cations stops because they have balanced all the OH^-, and the membrane becomes permeable to anions, the OH^- and the anions from outside would exchange (Fig. 8.2c). The net result would be that both ions of the salt would be accumulated *against* the concentration gradient. At the time each ion is being transported, it is

137

moving *with* the electrochemical potential gradient set up by the H^+ and OH^- resulting from the charge separation but, at the end, water is the only substance increasing outside the membrane. Since water is not measured in experiments measuring active transport, the mechanism would be difficult to explain until the role of the proton gradient was understood.

The inside of a cell or an organelle obviously contains many ions of different kinds and the consequences of proton gradients will be much more complicated than the simple hypothetical system we have considered. It was not until the early sixties that we began to appreciate what was happening in ion transport across the inner mitochondrial membrane. For some time it was observed that divalent cations, in the presence of phosphate, are absorbed by mitochondria. This absorption occurs in the presence of

Fig. 8.2. Possible redistribution of ions resulting from a proton gradient: (*a*) charge separation results in H^+ on one side and OH^- on the other side of a membrane; (*b*) if the membrane is cation permeable, cations can enter to balance the OH^-; (*c*) if the membrane changes from cation permeable to anion permeable, anions can exchange with OH^-; (*d*) the net result is accumulation of cations and anions on one side and water formation on the other.

(*a*)

(*c*) Membrane anion permeable

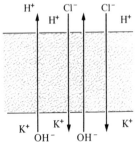

(*b*) Membrane cation permeable

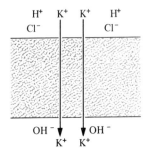

(*d*)

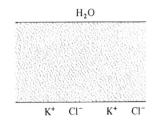

substrate supporting oxygen uptake, provided that ATP formation is blocked, either by omitting ADP from the medium or by adding oligomycin, the inhibitor of ATPase. Hydrogen ions continue to be secreted simultaneously. It is clear now that the uptake of cations was due to the electrochemical potential gradient established by the OH^- remaining inside the mitochondrion. At the same time, some OH^- exchanged for phosphate (Fig. 6.7). Calcium ions inside, with phosphate in an alkaline medium, result in the precipitation of calcium phosphate which can be observed with the electron microscope. The same effect was obtained in the absence of oxygen uptake by supplying ATP to the mitochondria; in these conditions, the hydrolysis of ATP, taken inside by exchange with ADP, results in extrusion of protons from the mitochondria and a rise in OH^- concentration in the matrix – in other words, the H^+/OH^- gradient results in the entry of other ions.

All the work with phosphate and divalent cations served to extend our understanding of the importance of the H^+/OH^- gradient, whether established by electron transport or by membrane ATPase hydrolysing ATP. By the mid-sixties most of the normal ion behaviour of mitochondria had been worked out. Three features are characteristic:

(1)   the inner membrane is not permeable to H^+ and OH^-;
(2)   the traffic in other ions depends primarily on the gradient of H^+ and OH^-;
(3)   specific carriers are responsible for the exchanges of ions.

Some of these properties have been shown in Fig. 6.7, but some further explanation is desirable here.

The work of Chappell & Crofts (1966) and Chappell & Haarhoff (1967) showed the sequence of ion movements which result from the gradient and the specific carriers, and our understanding has developed greatly since then. An antiport allows OH^- on the inside of the mitochondrion to exchange for an ion like phosphate (arsenite will also exchange), but not for chloride, bromide or sulphate. Other antiports allow certain dicarboxylic acids, e.g. malate, succinate, to enter in exchange for HPO_4^{2-} or in exchange for each other (Fig. 6.7). Thus, both the presence of phosphate and specific molecular conformations – the antiports – in the membrane are required. Other organic acids, citrate and cis-aconitate, enter only when the phosphate is accompanied by L-malate in low concentrations, but its effect is completely inhibited by butyl malonate. We see, therefore, that the ion entry mechanisms, for which the energy was derived from the respiratory charge separation, depend also on specific carriers in the membrane. Though the phosphate carrier and the ADP/ATP exchange systems are present in all mitochondria, the other carriers differ in different species. The presence or absence of the particular protein for a permease is genetically determined. Membrane proteins are the controlling factor in this specificity but the supplementary role of the lipids is not yet understood.

With all this happening, along with electron transport and other reactions not specified here, the inner mitochondrial membrane can be seen to be a very lively membrane.

ATP-dependent H^+/OH^- gradient

I have described ATP as an important energy currency of the cell, used in places other than the mitochondria or bacterial cell membranes in which it is formed. Some of these uses are in ion transport by the action of a specialised ATPase in the membrane concerned. Some of these transport systems depend on the formation of another H^+ and OH^- separation across the membrane. An example which is now well understood is in the chromaffin granules which occur in the cells of the adrenal glands (Johnson, Carlson & Scarpa, 1978; Johnson & Scarpa, 1979). These cells secrete the hormones norepinephrine (noradrenalin) and epinephrine (adrenalin) which are catecholamines. The amino group behaves as a base

Fig. 8.3. (a) The membrane of a chromaffin granule contains a proton-translocating ATPase which transports H^+ into the granule; $R-NH_2$ (catecholamine) and Cl^- can both pass through the membrane; $R-NH_2$ becomes protonated and $R-NH_3^+$ cannot pass back through the membrane; Cl^- balances the charge. (b) The sequence in granule behaviour: accumulation of catecholamine is followed by water uptake, swelling, fusion with the plasmamembrane and exocytosis.

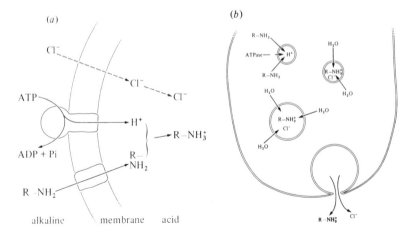

and therefore adds a proton, becoming positively charged at lower pH values. The chromaffin granules are small organelles surrounded by a membrane which is permeable to catecholamines at higher pH when they are not charged (Fig. 8.3). The membrane also contains an ATPase which hydrolyses ATP on the outside of the granule and produces H^+ ions inside. The H^+ ions inside are joined by a Cl^- ion which crosses the membrane. As the pH drops, any catecholamine which has diffused in as a neutral molecule becomes protonated and therefore positively charged. This positive charge is balanced by the Cl^-. The charged catecholamine cannot penetrate the granule's bilayer, so it remains trapped inside and more neutral molecules will enter to be similarly trapped by the steady stream of hydrogen ions from the ATPase. The accumulating ions inside increase the osmotic pressure of each granule which swells by uptake of water. Eventually the membrane of the granule makes contact with the cell membrane and fuses with it, but in such a way as to discharge its contents to the exterior of the cell. This process of membrane fusion is known as *exocytosis*, a common membrane phenomenon in absorbing or secreting cells.

This beautiful system of an ATP hydrolysis setting up an H^+/OH^- gradient leading to movement of specific substances across a membrane and later to their secretion, reminds us that ATPases are sometimes means of getting the protons where they are required for ion transport. The 'pumping' of adrenalin has become a favourite cliché for sports commentators describing almost every physical activity. Perhaps they would enjoy knowing how the pump works and change the cliché to active ATP hydrolysis!

Ion transport in bacterial cells

The same principles that we have been discussing in mitochondrial membranes apply to the membranes of bacterial cells. There, we can have a proton gradient established either directly by respiration (see Chapter 6) or indirectly by an ATPase in the membrane which will hydrolyse ATP, secrete H^+ ions externally and result in negative ions inside the cell. This ATP may have been generated by fermentation but, used by a membrane-bound ATPase, it can set up a proton gradient which will have the effect of redistributing ions. If there are specific carriers in the bacterial membrane, the redistribution will follow the same principles. Considerable evidence points to the entry of undissociated organic acids, the role of the proton outside the cell being to form the molecule which can then combine with the membrane carrier. If the pH is high inside the cell, the molecule dissociates and the ion will be unable to diffuse back through the membrane. More puzzling are the uptakes of neutral molecules like galactose or arabinose which are accompanied by protons on entry. One hypothesis is that the proton is necessary to neutralise an anion on the membrane carrier to allow the galactose to combine by hydrogen bonding.

Active transport not dependent on proton gradients

When it was found that active transport processes required energy to move ions against the electrochemical potential gradient, it was inevitable that the word *pump* would be introduced and come into common use. Convenient as this short word is to designate an active transport, it does not describe it. Indeed we must answer the question of how the molecular mechanism works and not be confused by looking for an analogy to pumps which are mechanical. The evidence for the presence of a pump can be easily obtained, particularly when we can measure two parameters:

(1) the electrochemical potential difference, for example between the inside and the outside of a cell,

(2) the flux of a particular ion, e.g. Na^+, K^+, Cl^-, especially easy when using isotopes of the ions being investigated. If the net flux of the ion is against its electrochemical potential, work is being done and a pump is said to be operating.

The Na^+/K^+ pump. A high concentration of K^+ and a low concentration of Na^+ compared to the surrounding medium is characteristic of most plant, animal and bacterial cells. This difference is maintained by a mechanism located in the cell membrane which results in K^+ being accumulated inside the cell and Na^+ being extruded. The mechanism is due to an ATPase which, in the course of hydrolysing ATP, puts K^+ inside the membrane and Na^+ outside, the Na^+/K^+ ATPase (Fig. 8.4). The enzyme is well understood in animal cells, particularly erythrocytes, and not so clearly understood in plant cells.

Characteristically, the function of the ATPase is dependent on both Na^+ and K^+ and also on Mg^{2+}. The Na^+ and ATP must be inside the membrane and the K^+ (or Rb^+ or NH_4^+) outside to activate the ATPase. When the ATP is hydrolysed, Na^+ is extruded and K^+ is absorbed. While this is going on, the ATPase itself becomes phosphorylated (clearly shown if $AT^{32}P$ is used because the radioactive ^{32}P is transferred to the enzyme). The phosphorylation requires Na^+ and Mg^{2+}; the dephosphorylation requires K^+. The phosphoryl group is attached to a glutamate residue of the enzyme protein. Despite our knowledge of this interesting membrane-bound protein, we still do not understand how the obligatory Na^+ and K^+ stimulated reactions are linked to their transfer outwards and inwards respectively. An appropriate change in the binding capacity of the active site on the membrane must be postulated but its nature awaits further knowledge of this protein. The sidedness of the reaction is illustrated by the fact that steroids, particularly ouabain, a plant product derived from *Digitalis*, the foxglove, inhibit the dephosphorylation by the potassium reaction, but only if the inhibitor is on the outside of the membrane. None of the reactions is possible if the enzyme is not in a suitable membrane, not only for the reaction to occur but also for the separation of the K^+ and Na^+ to be maintained.

The tight coupling of the reactions involved and the conservation of energy from ATP as a Na^+/K^+ gradient is shown by the way the reaction can be reversed. Garrahan & Glynn (1966) showed that $AT^{32}P$ could be

Fig. 8.4. Steps in membrane Na^+/K^+ pump activity: (*a*) $3Na^+$ on the inside of the cell stimulate combination of ATP with Mg^{2+} and the pump protein; (*b*) $2K^+$ from outside the cell enter and stimulate dephosphorylation; (*c*) $3Na^+$ leave to the outside and $2K^+$ and phosphate (Pi) leave to the inside.

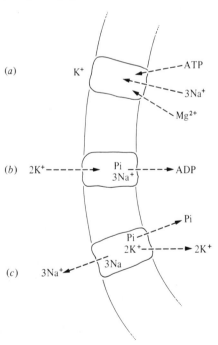

143

Fig. 8.5. The Ca^{2+} ATPase. (a) In the resting state the sarcoplasm of muscle accumulates Ca^{2+} in the cisternae due to the activity of the Ca^{2+} ATPase; (b) the ATPase phosphorylates the protein, stimulated by Ca^{2+} and Mg^{2+}; (c) Ca^{2+} is tightly bound to the protein; (d) Ca^{2+} and phosphate leave to the cisternal side of the membrane.

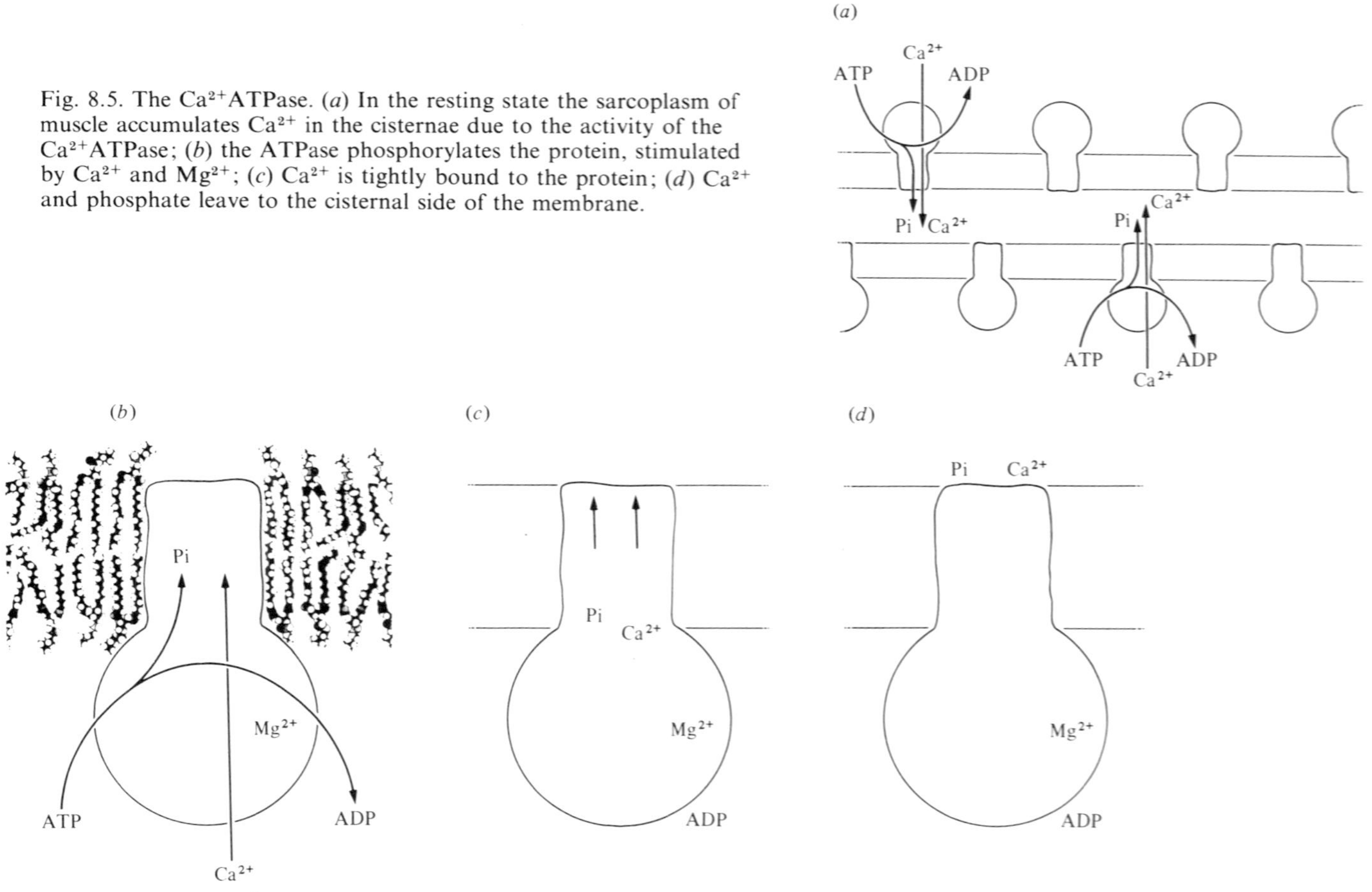

formed by the intact membrane of the red cell in the presence of potassium inside and sodium outside.

The Ca^{2+}ATPase. Another pump which depends on ATP is the *Ca^{2+}-ATPase*, which occurs in the membranes of the endoplasmic (sarcoplasmic) reticulum of muscle (MacLennan & Holland, 1975; MacLennan & Campbell, 1979). When muscles are at rest Ca^{2+} is stored inside the reticulum; when a nerve impulse activates the reticulum, Ca^{2+} is released into the vicinity of the muscle fibres and stimulates their contraction. The transport of Ca^{2+} through the sarcoplasmic reticulum membrane during the resting phase is brought about by the Ca^{2+}ATPase which depends on Ca^{2+}, ATP and Mg^{2+}. During the hydrolysis of ATP, phosphate is transferred to the enzyme under the influence of Ca^{2+} and Mg^{2+} (Fig. 8.5). The Ca^{2+} is also bound with high affinity and, at the end of the reaction, has been moved from the outside to the inside of the endoplasmic reticulum. The membrane is highly specialised as about 90% of the protein consists of the Ca^{2+}ATPase. Two other calcium-binding proteins also occur. If the phospholipids of the membrane are removed with phospholipase enzymes, the ATPase activity is lost but can be restored by adding back the phospholipids, about 30 of which form an annulus around the ATPase. As with the Na^+/K^+ ATPase, we do not understand the molecular changes involved in the protein to make this possible.

Transport dependent on ion gradients. If a cell or organelle has stored energy in the form of an ion or salt gradient, it is theoretically possible to have a reverse system so that the energy of diffusion to equilibrium can be brought back to use in ATP formation. An example of this is the reversal of the Na^+/K^+ pump already mentioned. However, a gradient of ions is sometimes used for other processes in transport. For example, the entry of sugars and amino acids into some animal cells depends on a gradient of Na^+ from the outside to the inside. Apparently Na^+ and glucose, for example, both bind to a transport protein in the membrane and enter the cell together. The Na^+ is subsequently extruded by the Na^+/K^+ pump. The molecular mechanism is not known.

Macromolecular secretion

Membranes are essential to cells which secrete large molecules with different functions, e.g. digestive enzymes from stomach and pancreas, the hormones, insulin from the pancreas and epinephrine (adrenalin) from the adrenal medulla, and cellulose for cell walls from plant cytoplasm. The

Fig. 8.6. The endoplasmic reticulum (ER) of chicken liver cells showing the numerous ribosomes on each side of the membranes and the cisternae between. M, mitochondrion. (Micrograph: J. M. Bain.)

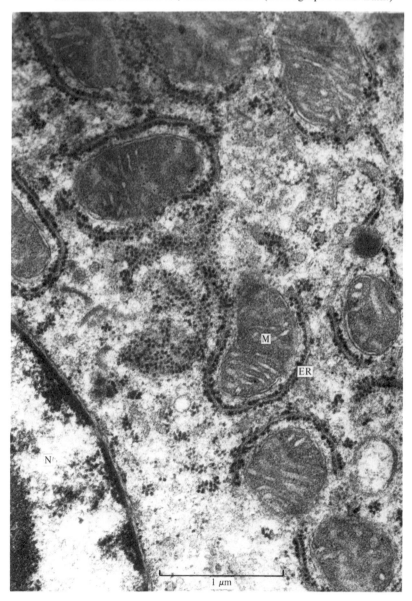

Fig. 8.7. Golgi body (GB) in *Euglena*, showing the stacks of membranes which bud off vesicles at their ends. M, mitochondrion; P, plastid. (Micrograph: J. M. Bain.)

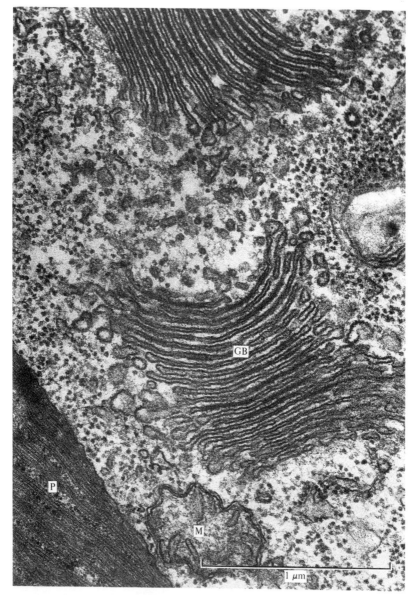

147

membranes which are involved are the endoplasmic reticulum, the Golgi apparatus, membranes of specialised vesicles and granules and the plasma membrane of the cell from which the secretion is discharged. An electron micrograph of the endoplasmic reticulum of chicken liver is shown in Fig. 8.6. This is termed *rough* endoplasmic reticulum because of its appearance which is due to the presence of many ribosomes adhering to the flat membranes on each side of the continuous compartments known as cisternae. We shall discuss the role of the membrane and the ribosomes when dealing with synthesis in Chapter 10. The other important system, the Golgi apparatus (Fig. 8.7), consists of flat membranes which often occur in stacks. The ends of the membranes in the stacks swell into vacuoles and bud off independent vesicles.

A good example of this kind of secretion is the way in which insulin, formed in specialised cells of the pancreas, is liberated into the blood

Fig. 8.8. Secretion of insulin in pancreas: (*a*) proinsulin (PI) is synthesised by ribosomes (R) attached to endoplasmic reticulum and pushed into the cisternae; (*b*) vesicles containing PI are budded off from the reticulum; (*c*) vesicles fuse with Golgi membranes and put PI between them by pinocytosis; (*d*) proteases of the Golgi membranes turn proinsulin into insulin (I); (*e*) vesicles are budded off Golgi membranes; (*f*) insulin in presence of Zn^{2+} crystallises in vesicles; (*g*) vesicles fuse with plasma membrane and, by exocytosis, extrude insulin into the blood.

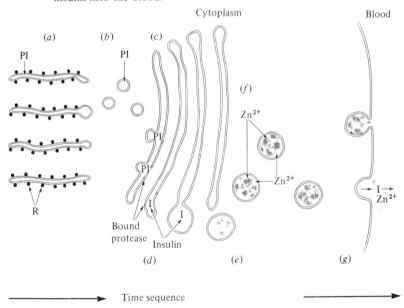

(Steiner *et al.*, 1974). The sequence starts in the endoplasmic reticulum where a polypeptide is synthesised under the influence of the ribosomes on the membranes (Fig. 8.8). The polypeptide, known as proinsulin, contains 84 amino-acid residues and is not active as a hormone. After a certain amount has been accumulated in the cisternae, smooth vesicles are apparently budded off and these, containing the polypeptide, pass to the Golgi apparatus where they fuse and, by pinocytosis, liberate the proinsulin in between the Golgi membranes. The membranes of the Golgi body contain bound proteases, enzymes which can split peptide linkages. The proteases split out a long part of the proinsulin chain, actually from amino acid 31 to amino acid 63 inclusive, leaving what had been the two ends of the molecule linked in two places by sulphydryl groups and thus forming insulin. The Golgi then proceeds to bud off vesicles, or granules as they are called, in which, under the influence of Zn^{2+}, both the insulin and the 33-amino-acid polypeptide crystallise out. The hormone remains stored in these granules and may be there for long periods, up to days. The next action takes place when the plasma membrane receives some signal that the blood-glucose content has risen above a certain level. When that happens, the granules nearest the plasma membrane fuse with it and by exocytosis the contents are ejected into the blood. The fusion of the membranes of granule and cell surface is dependent on Ca^{2+}. With our knowledge of the effects of calcium on adjacent negatively charged membrane molecules, we can expect that it will possibly contribute to fusion by pulling together anionic regions in the two membranes and starting the reorientation.

This example of secretion shows how several membranes are involved in sequence. Not only do they have specialised functions but also a coordinated sequence, all dependent on their very lively, varied activities.

Suggested reading

Finean, J. B., Coleman, R. & Michell, R. H. (1978). *Membranes and their Cellular Functions*, 2nd edn. Oxford: Blackwell Scientific Publications. (Chapter 4).

Harold, F. M. (1977). Membranes and energy transduction in bacteria. *Current Topics in Bioenergetics*, **6**, 83–149.

Harris, E. J. (1972). *Transport and Accumulation in Biological Systems*. London: Butterworth.

Quinn, P. J. (1976). *The Molecular Biology of Cell Membranes*. London & Basingstoke: The Macmillan Press Ltd. (Chapter 4).

9

Excited membranes and signal transmission

We have seen how a membrane with its insulating properties can establish and maintain an electrical potential difference. As long as the potential is maintained, it can be used as a source of energy for other processes. Somewhere, some time early in the evolution of animal life, this capacity to maintain an electrical potential became adapted to the vitally important function of conducting signals rapidly from one part of an organism to another. The modified cell which carries out this function in animals is the *nerve cell* or *neuron* and the mechanism is the conduction of an electrical impulse along the length of the membrane at the surface of the cell. Because the electrical impulse, which involves change in potential, can produce action at the other end of the cell, the impulse is called an *action potential* and cells which can produce such an impulse in their surface membranes are said to be *excitable*. Action potentials are not unique to animals; they also occur in plants such as the giant algal cells, where an electrical impulse is transmitted from one end of the cell to the other. In the sensitive plant, *Mimosa pudica*, which closes its leaflets together when it is touched, the message to close is transmitted through the plant by an action potential in some of the cells of the phloem.

However, it is in animals that the membrane mechanism to transmit messages reached its highest development. Without this rapid transmission of signals, animal life as we know it could not have evolved. Though the rates of transmission (up to $1 \text{ m}/10^{-2} \text{ s}$) are slow compared with the conduction of electricity by a wire ($1 \text{ m}/10^{-7} \text{ s}$), they are far faster than the rates of messages relayed by hormones circulated from one part of the body to another. Rates of transmission vary with different nerves and can be as low as about 1 m/s or as high as 100 m/s. Often the distances are quite microscopic but the messages are vital. Brain cells, for example,

150

communicate with each other by action potentials. It has been estimated
that human brain contains about 10^{10} neurons, so the electrical signal
traffic over short distances is intense.

In this chapter we shall be concerned with the way in which electrical
potentials *across* nerve cell membranes are set up and maintained and how
the electrical signal is transmitted *along* the membrane. We shall see that
different properties are necessary at nerve endings or *synapses* where the
message is transmitted from one nerve to another or from nerves to other
cells, e.g. muscle cells which become activated at the special synapses also
called *neuromuscular junctions.*

Neurons

Cell structure

Though the neurons vary greatly in size and shape, all have
essentially three parts (Fig. 9.1):

(*a*)   the *cell body* containing the nucleus;

Fig. 9.1. Diagram of a nerve cell: (*a*) nucleus; (*b*) dendrites; (*c*) axon;
(*d*) branches of axon; (*e*) synapses.

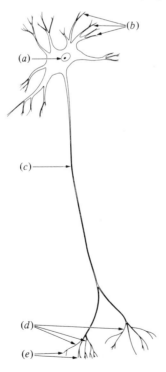

151

(b) *dendrites*, which are branched extensions of the cell surface, and

(c) the single *axon*, which is often a very long extension, up to about 1 m in length.

Axons are primarily concerned with the transmission of the electrical signal and make contact at their ends, either with other neurons or with effector organs, e.g. muscle cells which, on receiving the electrical signals, are responsible for action. For our purposes we shall attempt to understand the properties of the axon membrane and of the membranes involved when a nerve ending makes effective contact with another cell. Axons are frequently branched where they join the other cell in the synapse (Fig. 9.2a). In some synapses, there is electrical conduction between the membrane bringing the electrical signal, the *presynaptic membrane*, and the membrane receiving the signal, the *postsynaptic membrane*. In other synapses, the presynaptic membrane and the postsynaptic membrane are separated by a small gap across which a chemical transmitter diffuses (Fig. 9.2b).

Fig. 9.2. (a) Nerve endings making contact with a cell which is passing on the signals; solid arrows, the signals received; dotted arrow, the signal sent on. (b) A nerve ending or synapse showing (i) the presynaptic membrane and (ii) the postsynaptic membrane with the gap between.

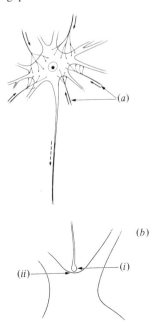

Membrane composition

The surface membranes of nerves and of other excitable cells are lipid bilayers, of the kind we have been discussing, accompanied by specialised proteins. One important protein in the axon of nerves is that of the sodium/potassium pump (Na^+/K^+ ATPase) but other essential proteins occur. Some of the proteins of the synaptic membranes are specialised for functions which do not take place in the axons.

Axon membranes

Details of the composition of axon membranes have been very difficult to obtain, especially as the membranes cannot easily be purified. Data on membrane fractions of first stellar nerves of squid, which were probably mostly axon plasma membranes, showed that lipids comprise about 70% and proteins about 30%. The commonest lipids were cholesterol and phosphatidylcholine, each about 27% of the total lipids, with phosphatidylethanolamine, about 20% next. Phosphatidylserine, which carries a net negative charge, and sphingomyelin were present in small amounts, 5% each. Also present were 4% of fatty acids and 9% of hydrocarbons (Camejo et al., 1969). We do not know how the different lipids are distributed between the two halves of the bilayer but the presence of substantial amounts of both cholesterol and phosphatidylethanolamine, which have shapes of inverted cones, suggests that, as in erythrocytes, they may occur in different layers or the packing of the cylindrical bilayer would not be possible (see Chapter 3).

In addition to the sodium/potassium pump other proteins are present. Some seem to be concerned with the sodium or potassium diffusion channels and will be discussed later.

Synaptic membranes

Not surprisingly the membranes in the synapse are different in composition from those in axons and have a higher percentage of proteins (50%). Analyses have been done on synaptic plasma membranes from rat brain (Cotman et al., 1969; Breckenridge, Gombos & Morgan, 1972). Phosphatidylcholine is the commonest lipid with 34% of the total. Phosphatidylethanolamine is the next most common with 28% and cholesterol, at only 19%, is relatively less important than it is in axons. Phosphatidylserine, with 10%, is relatively more important than it is in axons. Phosphatidylinositol is present at 2% and sphingomyelin at 3%. The hydrocarbon chains of phosphatidylcholine and phosphatidylethanolamine contain a large proportion of long-chain, unsaturated fatty acids.

Though phosphatidylethanolamine has 23.7% of C_{18} side chains with no double bonds, it has 18% consisting of C_{20} side chains with 4 double bonds and 32.9% of C_{22} chains with 6 double bonds. Most of the phosphatidyl-choline has chains of C_{16} or C_{18} with no double bonds (63.3%), some C_{18} with one double bond (24.2%) and smaller amounts of C_{20} with 4 double bonds (5.6%) and of C_{22} with 6 double bonds (3.4%). The phosphatidyl-serine, though having 48.6% of C_{18} chains with no double bonds, has 34.1% of C_{22} chains with 6 double bonds. Phosphatidylinositol, 34.8% C_{18} with no double bonds, also contains 36.9% of C_{20} with 4 double bonds. Sphingomyelin contrasts with the other lipids because it has 87.5% saturated C_{18} chains.

As we know from earlier discussions, the predominance of the double bonded hydrocarbon chains, together with low amounts of the rigid cholesterol, will confer great fluidity on the membranes of the synapse. This fluidity may be very important to their functions which will be discussed later.

The sodium/potassium pump occurs in the membranes of the synapse and maintains a potential difference between inside and outside. The proteins associated with the sodium and potassium diffusions also occur in the synapse, as well as some other polypeptides whose functions are not yet understood. Other important proteins occur in the membrane of the cell receiving signals across the synaptic gap and will be discussed later.

The resting potential

The electrical potential maintained across the surface membrane of the nerve cell is due primarily to the difference between the sodium and potassium concentrations on the two sides of the membrane, maintained by the Na^+/K^+ pump. Our understanding of the importance of this potential and of its nature owes much to the observation that the axon of the squid is very large (up to 1 mm in diameter) and it was therefore possible for Hodgkin & Huxley (1945) to record the action potential of this

Table 9.1. *Ionic composition, squid axon*

	Axoplasm mmol/l	Blood mmol/l
K	410	22
Na	49	440
Cl	40	560

154

axon with an electrode inside the cell, and for Hodgkin & Katz (1948) to measure the resting potential. However, very rapid developments with microelectrode techniques which allowed electrodes to be inserted into nerve and muscle fibres, soon made it possible to measure membrane potentials from the inside to the outside of nerves in many organisms. The potential difference between the inside of the axon and the blood or body fluid on the outside is related to the differences in ions on the two sides of the membrane. Table 9.1 taken from Hodgkin (1951) shows the substantial difference, particularly that potassium is much higher in the axoplasm than outside and that sodium and chloride concentrations are much lower inside than out. The potential difference between the two sides is usually between 50 and 90 mV, negative inside. The difference in the resting cell, and hence the resting potential, is maintained by the sodium/ potassium pump, as shown by Hodgkin & Keynes (1955), which continually extrudes Na^+ and brings in K^+. Since the membrane is not perfectly impermeable to K^+ and to Na^+, the pump which, as we have described earlier, derives its energy from the hydrolysis of ATP, must work all the time. One potassium enters for every two or three sodiums pumped out (Fig. 9.3). Such pumps operate in all known excitable cells to maintain the resting potential.

The action potential

Once the electrode techniques for determining the potential between inside and outside the cell were perfected, it was possible to be more precise about the electrical change which accompanies the action

Fig. 9.3. The resting potential is maintained by the sodium/potassium pump (*a*) which extrudes Na^+ and brings in K^+, thereby working against (*b*) the tendency for Na^+ to diffuse in and for K^+ to diffuse out; (*c*) other membrane proteins are present.

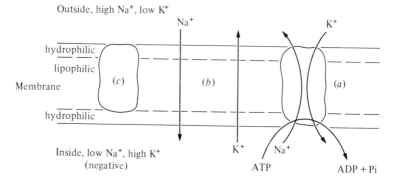

155

Fig. 9.4. A diagrammatic representation of the action potential passing along the axon's surface membrane as observed between an external and an internal electrode: (*a*) the change in potential across the membrane approaches the electrodes but the potential remains at resting value (− 50 mV); (*b*) the electrodes show the transient change in potential as it goes from − 50 mV to a small positive value and returns to − 50 mV in about 1 ms; (*c*) the signal has passed and the resting potential is restored.

(*a*)

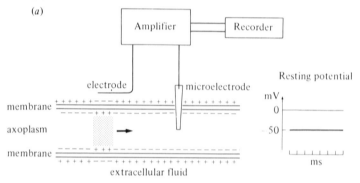

(*b*)

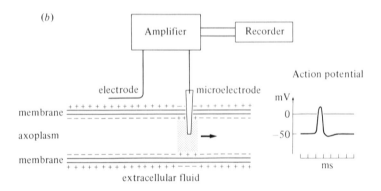

(*c*)

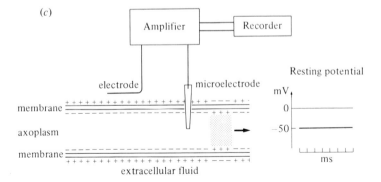

potential, than from early observations made with two electrodes on the outside of the cell. It soon became clear that an action potential is really a transient change in sign of the potential across a surface membrane. When a nerve goes from a resting potential to an action potential, measurements at any one spot on the membrane show that the potential difference between inside and outside is quickly reversed but then just as quickly returns to the resting state when the impulse is past (Fig. 9.4). All this happens in about one millisecond (10^{-3} s). The membrane in the resting state is negative inside.

With some very clever experimental techniques, Hodgkin & Huxley (1952) were able to sort out the current components of the action potential, in classical work which led to the award of the Nobel Prize in Physiology in 1963. The early inward current is associated with the inward diffusion of sodium ions, to which the membrane suddenly becomes permeable, so positive charge is carried in. This sets up a reversed potential across the membrane but the sodium channels then close so potassium channels allowing an outward diffusion of potassium restore the inside negative potential again. It is a simple and probably correct postulate that the flow of sodium inwards in one area of the membrane causes membrane depolarisation which then affects the permeability in the adjacent area so more sodium enters and affects the next adjacent area, continuing the effect right along the membrane.

At present our knowledge of what components in the membrane are involved in these separate Na^+ and K^+ channels is somewhat limited. Why should a channel which opens to allow sodium to enter not be penetrated by potassium which would leave at the same time? One possibility is that the sodium channel can allow the hydrated sodium to pass whereas the hydrated potassium would be too big. Sodium could pass through a channel with a cross section of 0.3×0.5 nm at its narrowest point but hydrated potassium could not. However, both small metal (lithium) and organic cations (hydroxylamine, H_3N^+—OH, and hydrazine, H_3N^+—NH_2) can pass through the channel so the explanation of permeability is more complicated than mere size and is probably related to the affinities of ions for binding sites in the channel compared with their affinities for water.

The infamous nerve poison tetrodotoxin, one of the deadliest poisons known, which comes from the puffer fish regarded in Japan as a delicacy provided all the poison has been removed, has been shown to block the sodium channel and thus to inhibit the inward current. Tetrodotoxin is a molecule with a molecular weight of 319 (Fig. 9.5a), which combines very effectively with the site of the sodium channel in the membrane but is much too large to pass through. It acts only on the outside of the membrane and is almost completely ineffective if applied to the inside. It seems that the positively charged guanidinium group of the molecule enters the channel while other groups become associated with adjacent reactive parts of the membrane, perhaps by hydrogen bonding. It is interesting to note that the binding of tetrodotoxin is unaffected by many enzymes, e.g. phospholipase C which hydrolyses the bond between phosphoric acid and glycerol, phospholipase D which removes the polar head group to leave only the phosphatidic acid; it is also unaffected by the proteolytic enzyme, trypsin, which attacks peptide bonds whose carbonyl function is donated by either a lysine or an arginine. However the binding property of tetrodotoxin is affected by treatment with chymotrypsin which attacks peptide bonds whose carbonyl function is provided by phenylalanine, tryptophan or tyrosine. It is particularly important that the binding function of tetrodo-toxin is also inhibited by phospholipase A which removes the fatty acid from one of the positions in the phospholipid. This would convert the diacyl phospholipids to a single acyl or lysophospholipid. We have seen in Chapter 3 how this shape (a cone) will not be suitable for maintaining a bilayer. It is reasonable to speculate that the packing of the bilayer molecules is completely altered by this change and thus the molecule (or molecules) of the Na^+ channel is no longer able to present the correct face

Fig. 9.5. (a) Tetrodotoxin which blocks the sodium channel. (b) Tetraethylammonium which blocks the potassium channel.

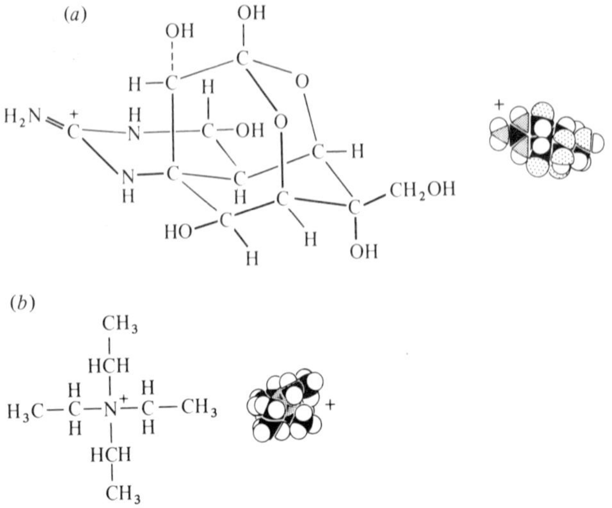

for the tetrodotoxin to bind. It is significant that phospholipase A inhibits action potentials and this is probably also due to an alteration in the packing of the molecules in the bilayer. In 1972, Benzer & Raftery concluded that the binding site for tetrodotoxin in the sodium channel was a protein and the protein has now been isolated by Agnew *et al.* (1978). This story is a fine illustration of the compulsory association of a functional protein with the correct and correctly packed lipids of the bilayer. Despite the advancement of our knowledge to this stage, as Keynes (1979) has said, any portrayal of the inner workings of the sodium channel must be strictly fanciful.

In 1972 Keynes pointed out that $1 \mu m^2$ of membrane contains about two million phospholipid molecules but only 13 to 75 Na^+ channels. The number of channels probably varies in different organisms. In squid axon, the potassium channels seem to be about 50 times more frequent than the sodium channels. Potassium channels can be blocked by quaternary ammonium compounds which are cations, e.g. tetraethylammonium, with some hydrophobic properties (Fig. 9.5*b*). It has been suggested (Armstrong, 1971) that the potassium channel is a gramicidin-like tunnel. Where the potassium enters on the inside of the membrane, a hydrated ion can enter but only a dehydrated K^+ goes through the pore. Whether this channel is entirely polypeptide or how far it is associated with adjacent essential lipids, is not known.

We have seen how the action potential depends on the diffusion first of sodium and second of potassium through channels which must open as the impulse arrives and close after it passes. The means by which these channels open and close, known as the *gating mechanism*, remains something of a mystery, though gating currents, i.e. the pulses of current generated when the gate opens, and charges move across or in the membrane, have now been measured. How does the action potential, arriving in adjacent parts of the membrane, open the channels, both of which start to open at the same time, but the potassium channel more slowly than that of sodium? The answer will lie in the electrochemical effects of the membrane complexes which may be on both protein and lipid constituents. The possible effects of Ca^{2+} ions and H^+ ions are also yet to be clarified.

Whatever the molecular details of the mechanism, this very impressive membrane organisation receives an electrical signal from a receptor in the membrane, which might have been activated by a chemical, electrical or mechanical stimulus, and transmits it along the axon. A receptor will be a protein which responds to stimulus with a conformational change which, in turn, sets off the nerve transmission. The receptor and the membrane potential must revert to their resting state with great rapidity so that the nerve cell is set for its next stimulus and the axon for its next transmission.

Synaptic transmission

In lower animals and in some tissues of vertebrates (e.g. heart muscle, cells of bladder wall), excited cells may have direct electrical contact at a synapse and transmit an action potential directly. However, in most vertebrate nervous systems, where one neuron ends, specialised membrane phenomena are associated with the transfer of the message to the membrane of the next cell, i.e. to the postsynaptic membrane. The postsynaptic membrane may belong to another nerve cell or to a muscle cell which will respond to the signal. Between the nerve ending and the postsynaptic membrane lies the synaptic gap, about 50 nm in width. Across the gap the communication is by a chemical substance – a chemical transmitter – which is a small diffusible molecule. Several kinds of molecules, viz. acetylcholine, epinephrine (adrenalin) and norepinephrine (noradrenalin) are effective as transmitters in different synapses. The structure of acetylcholine is illustrated in Fig. 9.6.

Acetylcholine is the transmitter substance between the nerve and the muscle membrane at the neuromuscular junction. Our knowledge of what happens will be reviewed here with particular reference to the membrane processes involved. The end of the axon which is bringing the nerve impulse is filled with many minute vesicles which are about 40 nm (Fig. 9.7a) in diameter and are filled with acetylcholine. The acetylcholine is synthesised in that part of the cell by the enzymatic transfer of the acetyl group from acetyl CoA to the base choline. Though about 50% of acetylcholine remains in the axoplasm, the rest is taken up by the vesicles each of which comes to contain about 6000 to 10000 molecules of acetylcholine. Assuming that the vesicles have a normal bilayer, they will, on a molecular basis, look something like Fig. 9.8. It is interesting to note that the research on

Fig. 9.6. Acetylcholine.

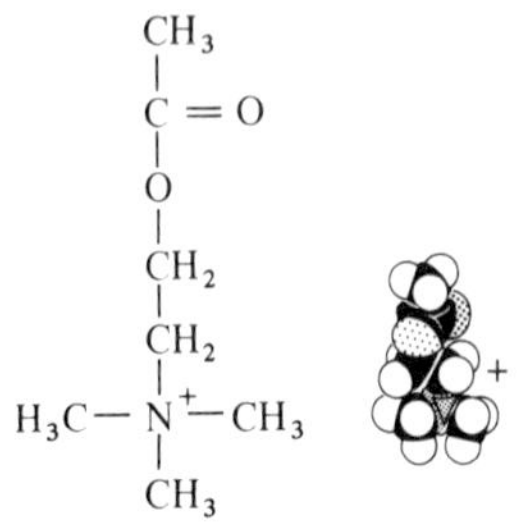

synaptic transmission predicted, before the vesicles were discovered, that there were definite packets or quanta of the chemical transmitter. They became definite entities when seen with the electron microscope and they can be extracted and purified by suitable techniques.

When a signal arrives at the nerve ending, the membrane is depolarised and acetylcholine is released into the gap (Fig. 9.7b). This happens when

Fig. 9.7. (a) A diagram of a synapse: (i) axon branch membrane, (ii) synaptic vesicle, (iii) synaptic cleft, (iv) postsynaptic membrane.
(b) (i) synaptic vesicle approaches presynaptic membrane as action potential arrives, (ii) the vesicle and presynaptic membranes fuse and the chemical transmitter crosses the synaptic cleft (iii) and triggers the action potential in the postsynaptic membrane (iv) and (v).

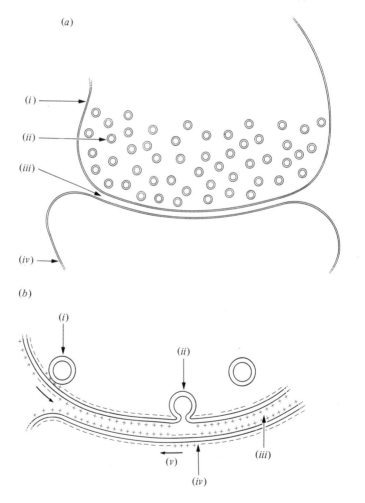

Fig. 9.8. An impression of a synaptic vesicle with acetylcholine dissolved in water enclosed by the vesicle membrane.

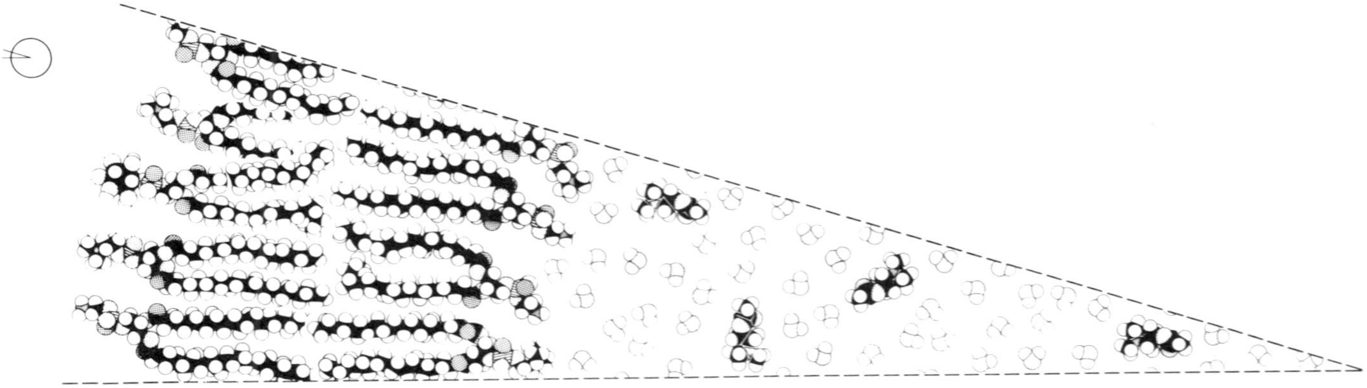

the synaptic vesicles, relatively stable in the axoplasm, are induced to fuse with the nerve-ending membrane and, in doing so, put the acetylcholine into the gap by exocytosis. This membrane behaviour would be consistent with some of the properties of lipid–protein bilayers which we have described. Though the change of charge due to the action potential may allow the vesicle to approach closer to the presynaptic plasma membrane, in the area which is temporarily positively charged, and the packing of some of the membrane molecules may be affected, it is not enough to cause fusion. Calcium ions are absolutely necessary for the fusion; Mg^{2+} will not substitute. Ample evidence shows that the calcium in the synaptic gap is essential to the release of acetylcholine and that it enters the cell when the action potential arrives at the nerve ending. Indeed, it can be made to enter the cell when the nerve-ending membrane is depolarised artificially with applied currents when the sodium channels are blocked with tetrodotoxin. Whatever else the calcium does on entering the cell, it will then be in contact with the synaptic vesicles and the nerve-ending membrane. Martin & Miledi (1978) found that calcium injected into the nerve ending caused synaptic vesicles to disappear and membrane invaginations to appear. It is tempting to speculate on the possibility that this calcium ion, acting between two negative charges on the adjacent membranes, will cause some rearrangement or different packing of the molecules, especially in a membrane which is so fluid due to the double bonds in the lipids. We have discussed how divalent cations can pull membrane molecules out of their alignment (cf. the reversal of phase in a soap emulsion) but fusion of pure lipid bilayers is not as easy as fusion of mixed lipid–protein membranes. The calcium may be acting with protein groups to facilitate the entry of synaptic vesicle membranes into the plasma membrane. There, the repacking which follows could result in the fusion, the incorporation into the planar membrane and hence in the liberation of acetylcholine into the gap.

Once the acetylcholine is released into the gap, it diffuses the short distance and combines with a receptor molecule in the postsynaptic membrane. This combination results in an increased permeability of the postsynaptic membrane to sodium and potassium, but not to chloride. The resultant depolarisation opens other sodium and potassium channels and an action potential results, to be transmitted along the membrane of the receiving cell in the usual manner. We do not understand the nature of the receptor for acetylcholine, though its probable molecular weight and subunit constituents are known. Since the acetylcholine is effective only if it is applied to the outside of the postsynaptic membrane, we can conclude that the active site of the receptor is to the outside. There is evidence that the bulk of the receptor protein is buried in the membrane.

The receptor is completely and irreversibly inhibited by the snake venom component, α-bungarotoxin, which combines with it and which, when labelled with tritium, can be used to estimate the number of receptors (about 10000 per μm^2). Once the acetylcholine has acted with the receptor, it is rapidly destroyed in hydrolysis to acetic acid and choline by the enzyme acetylcholine esterase which also occurs in the postsynaptic membrane with a frequency of about 2500 sites per μm^2. The postsynaptic membrane is then set to receive the next signal in the form of acetylcholine.

Anaesthetics

In discussing bilayers, I referred to the way in which small lipid-soluble molecules can affect the membranes by dissolving in the lipophilic region and altering the dimensions of the bilayer (Chapter 3). For a long time it has been known that anaesthesia, which is caused by a large variety of substances with small molecules, depends on their solubility in lipids. The local anaesthetic, benzyl alcohol has been shown by Ashcroft, Coster & Smith (1977) to increase the thickness, which would loosen the packing, of a bilayer. Much work in this field has been done by Bangham and his associates (Bangham, Hill & Mason, 1980). Based on the experiments they have done with liposomes, they suggest that anaesthetics can sufficiently interfere with the packing properties of nerve membranes to make them incapable of carrying out their normal functions of transmitting signals. That the disruption might be due to the small molecules loosening up the membrane constituents is suggested by the very interesting experiments with newts or tadpoles anaesthetised with chloroform or ether. When anaesthetised, they cease swimming and sink to the bottom of their tank. If, however, a pressure of about 90 atmospheres is applied to the tank, the newts or tadpoles resume their swimming (see Fifield, 1980). It is suggested that the high pressure restores order in the membranes of the animals in the same way as Bangham has shown that liposomes, which become leaky when an anaesthetic is applied, will cease to leak if put under a pressure of about 90 atmospheres.

Just how the anaesthetics act on nerve membranes to prevent their normal functions is not known. As we have seen, nerve membrane activities depend on coordination of membrane structures, both in the conduction along the axon and at the synapse; any alteration of the spacing of the relevant proteins or increase in the leakiness of the membranes, or both, could upset these essential interactions. We can expect considerable advances in our understanding of anaesthesia as the membranes are investigated further.

In this chapter we have discussed one of the most complex and beautifully coordinated membrane systems known. In living organisms a variety of stimuli – light, heat, mechanical (as in injury), chemical (as in smell and taste) – acting on the appropriate specific receptor, sends an electrical signal in the form of a nerve impulse to the brain. There, the cells also communicating with each other by action potentials have the whole complicated circuitry which is the basis of mind, including memory, all dependent on membrane phenomena. It is interesting to note that the short-term memory is dependent on sodium/potassium balance in brain cells. Mark (1979) has shown that, if the sodium/potassium pump is blocked by the specific inhibitor, ouabain, chickens which have been trained to follow simple routines for rewards, cannot remember what to do. Since ouabain is used with beneficial effects in the treatment of some patients with coronary heart troubles, humans, whose sodium/potassium pumps in the brain are also inhibited by ouabain, report confusion, which is loss of short-term memory (personal communication, Huppert & Tucker).

Suggested reading

Cotman, C. W. & Levy, W. B. (1975). Membranes in nerve impulse conduction. In *Biochemistry of Cell Walls and Membranes. Biochemistry* Series I, vol. 2, ed. C. F. Fox, pp. 187–205. London, Baltimore: Butterworths University Park Press.

Junge, D. (1981). *Nerve and Muscle Excitation*, 2nd edn. Sunderland, Massachusetts: Sinauer Associates Inc.

Keynes, R. D. (1979). Ion channels in the nerve-cell membrane. *Scientific American*, **240**, No. 3, March, 98–107.

10

Membrane-bound reactions – hormones, antibodies and synthesis

In this chapter we shall be concerned with more membrane-bound reactions, some of which control the metabolism of cells and organelles. We have seen how membranes restrict the entry of unwanted substances or allow the entry of others which are essential. In the last chapter we saw how messages are transmitted by the surface membranes of nerve cells. Here we shall discuss how another kind of message is received on the outside of the cell membrane and transmitted across to initiate essential reactions in the cell. For example, some hormones arriving at a cell surface do not penetrate the cell but trigger reactions inside. Similarly antigens at the cell surface produce messages which cross the membrane and divert synthesis in the cell to antibody production. Finally we shall see how membranes themselves are engaged in the synthesis of their own constituents, lipids and proteins, as well as in the production of proteins for export from the cell.

Hormones

The chemical messengers of animals, hormones, are carried by the blood stream from where they are formed to the target organs in which they regulate metabolic and physiological functions. Some hormones, e.g. sex and adrenal-cortex hormones are lipid-soluble steroids and can, as we would expect, pass through the membranes and come into contact with receptors within the cells. However, other hormones, e.g. epinephrine (adrenalin), glucagon and insulin, which are water-soluble, do not enter the cells but react with receptors in the cell membrane. Our particular interest is to consider how the membrane receives the hormone molecules on its specific receptors and how the message is transmitted across to trigger reactions in the cell. A number of hormones which sit on the outside of the cell produce the same effect inside, i.e. they stimulate the production of cyclic adenosine monophosphate (cyclic AMP).

The effects of the hormone epinephrine (adrenalin) which, as we have seen (Chapter 8), is secreted by the chromaffin granules in the cells of the renal medulla, is an appropriate example. After travelling in the blood stream, the epinephrine, arriving at the surfaces of various organs, results in increased blood pressure, heart rate and other muscle effects. With liver cells, it stimulates the breakdown of glycogen which results in an increase in blood glucose, but the hormone does not enter the cell. When it combines with the specific receptor on the *outside* of the cell membrane, a membrane-bound enzyme (adenylate cyclase), converts ATP *inside* the cell to cyclic AMP and phosphate (Cuatrecasas, 1974). This sets off a series of other reactions which result ultimately in the liberation of glucose (from the breakdown of glycogen) into the blood (Fig. 10.1). The cyclic AMP formed removes the regulatory unit from a protein kinase which, then being activated, initiates the other well known reactions, which need not concern us here. The hormone also stimulates (via the formation of cyclic AMP) glycogen synthesis in the liver.

As long as epinephrine is present outside the cells, the liver system remains activated, maintaining cyclic AMP at high concentration. When the hormone is no longer secreted by the renal medulla and therefore the concentration in the blood falls, the receptors and epinephrine dissociate, cyclic AMP is no longer formed and the other intracellular reactions cease. How the small epinephrine molecule (Fig. 10.2a) on the outside of the membrane activates the adenylate cyclase, which causes the formation of cyclic AMP on the inside, remains mysterious. Presumably the membrane-bound enzyme undergoes some kind of conformational change which, on the inside of the membrane, results in an active site combining with ATP and splitting off the phosphates. How far the epinephrine penetrates the membrane to have this effect is unknown.

Another hormone, glucagon, secreted by the pancreas, induces an effect in liver cells similar to that of epinephrine. But glucagon differs markedly

Fig. 10.1. Epinephrine stimulates liver cells to convert glycogen to glucose: (*a*) receptor protein and adenylate cyclase in the membrane; (*b*) epinephrine (epn), attaching to receptor, signals (perhaps by conformational change) adenylate cyclase to become active and cyclic AMP is formed; (*c*) cyclic AMP activates protein kinase; (*d*) other reactions follow and result in glycogen breakdown to glucose; (*e*) glucose passes through membrane to blood.

167

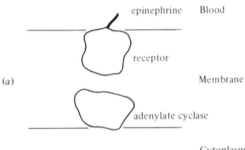

(a)

Blood

epinephrine

receptor

Membrane

adenylate cyclase

Cytoplasm

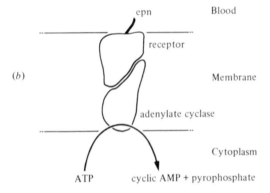

(b)

Blood

epn

receptor

Membrane

adenylate cyclase

Cytoplasm

ATP   cyclic AMP + pyrophosphate

(c)   cAMP + inactive protein kinase ⟶ cAMP–R + active protein kinase
           (PrR)   (Pr)

(d)   other reactions result in glycogen breakdown to glucose

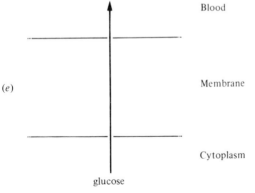

(e)

Blood

Membrane

Cytoplasm

glucose

from epinephrine because it is a polypeptide containing 29 amino-acid residues. Unlike epinephrine it has no action on other cells.

Other hormones are also known to increase the cyclic AMP in the cells on which they act but each has its own specific receptor in the cell surface. The subsequent reactions of cyclic AMP in these different cells are not as well understood as those in liver. With our knowledge of membranes, it is interesting to note that cyclic AMP action may be markedly affected by Ca^{2+} and that prostaglandin E_1 (Fig. 10.2b), which will tend to be membrane-bound also has a marked effect. Prostaglandin E_1 has been suggested to serve as an intermediate between the cell surface receptor and the membrane-bound adenylate cyclase in some cells. These speculations emphasise our ignorance about what might be happening in membrane-bound enzymes and the conformational changes which occur in them but also emphasise our need to understand what is happening in the adjoining lipids of the bilayer.

Fig. 10.2. (a) Epinephrine. (b) Prostaglandin E_1.

(a)

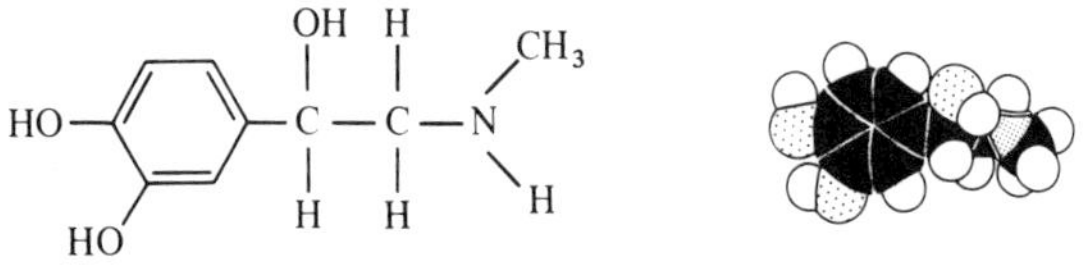

(b)

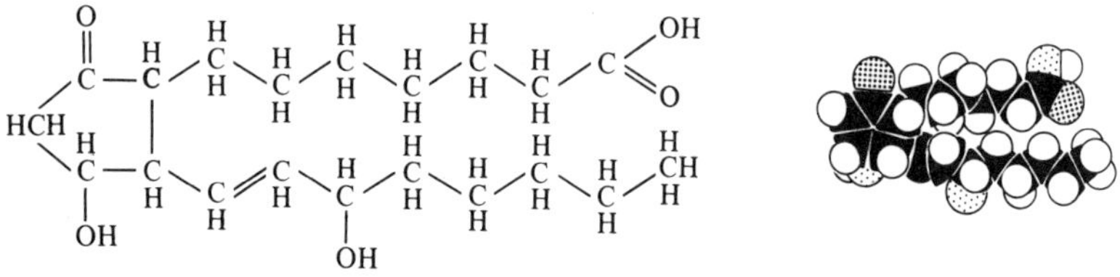

Cell recognition, antigens and antibodies

Surface membranes are obviously of great importance in controlling the effect of the chemical environment on the cell. Sometimes these membranes, as in the mammalian plasma membrane, are in direct contact with the external solution; at other times, as with the plasmalemma of plants or the surface membranes of some bacteria, they are protected by a cell wall. In the vertebrates, they are the sites of the cell-specific antigens, of the receptors for external antigens and of the receptors which recognise foreign bodies or accept 'self' bodies as with adjacent cells. They are partly also responsible for the adhesion between cells.

In Chapter 2 we saw that membrane-bound molecules include the glycosphingolipids. The simplest of these are the cerebrosides which have one monosaccharide molecule bound to the hydroxyl group of the ceramide (Fig. 2.9). Some have disaccharides or other oligosaccharides bound to their head groups; these are neutral lipids and some, found in the surface of red blood cells, are responsible for blood group specificity, i.e. they are antigens and can be detected because when they bind to their specific antibodies, which have two combining sites each, cells are held together and agglutination results (Fig. 10.3). Other glycosphingolipids (gangliosides), which have a single negative charge due to the presence of

Fig. 10.3. Agglutination of blood cells is due to the presence of antigens in the cell surface, each of which can combine with a binding site on an antibody molecule. Since antibodies have more than one combining site they act as bridges between the cells and the mass agglutinates. (Not to scale.)

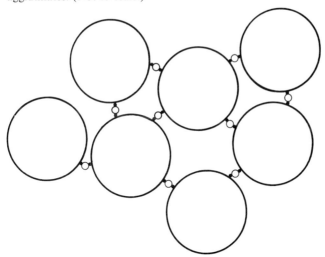

an acidic sugar (N-acetylneuraminic acid or N-glycosylneuraminic acid) in the oligosaccharide head group, also occur in surface membranes (Fig. 10.4).

Some of the proteins which occur in cell surfaces also have carbohydrate residues. These glycoproteins are known to be receptors for antigens. The oligosaccharide chains of the glycoproteins are restricted to the outsides of the cells where they can react with the foreign substances or, alternatively, recognise a compatible substance in another cell. Such proteins play roles in blood group specificity, in recognition and rejection of unlike cells, e.g. in grafts, as receptor sites for the influenza virus antigen and for the substance, phytohaemaglutinin. Lectins, which are plant glycoproteins that combine specifically with a carbohydrate of the cell surface (glycolipids or glycoproteins), are commonly used to investigate the cell surface receptors. The two glycoproteins, concanavalin A (which comes from jack bean) and wheat germ agglutinin, combine with surface receptors and alter the properties of the membrane.

This is not the place to engage in a discussion of the many fascinating aspects of the immune response. From the membrane point of view, it is interesting that an antigen bound to a receptor in the cell surface

Fig. 10.4. The constituents of a ganglioside: (a) model of galactose to compare the size with (b) N-acetylneuraminate and (c) the diagrammatic structure of the ganglioside (gal, galactose; gal Nac, N-acetylgalactosamine; Nan, N-acetylneuraminate; gl, glucose; cer, ceramide).

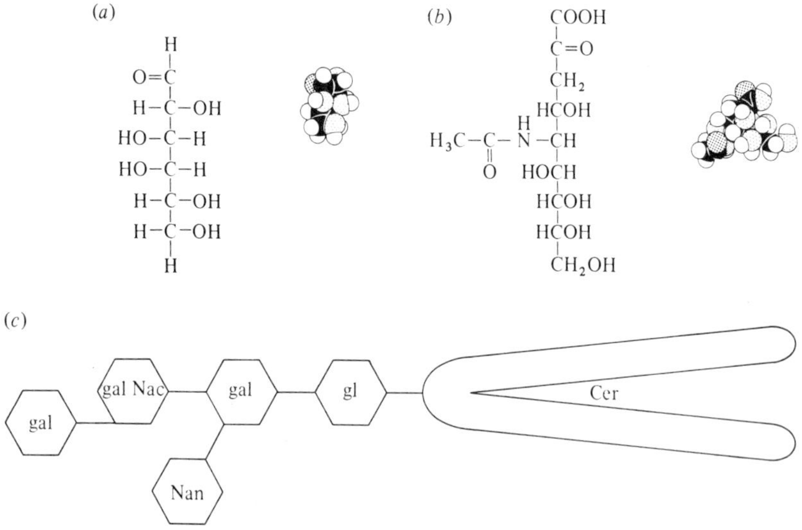

membrane can initiate changes in some cells to increase production of antibody molecules, proteins known as immunoglobulins which are quite specific in combining with the antigen. The cells, known as lymphocytes, synthesise antibodies and are stimulated to divide to produce plasma cells which continue to manufacture antibodies.

Lymphocytes show an interesting membrane phenomenon when they come into contact with an antigen; antibody molecules, which had been randomly spread round the surface of the cell, come together accompanied by their bound antigens at one area of the cell surface and form what is called a cap because of its appearance in the microscope. This fusion of the antibody–antigen complexes in the cell surface appears to be partly due to the antigens' ability to cross-link the antibody molecules. After this, the antibody–antigen molecules pass to the interior of the cell by a kind of pinocytosis. How this sequence occurs and how, on entering the cell, the complex stimulates the lymphocyte to further antibody synthesis is not fully understood.

Synthesis

Since membranes act as sites for enzyme activities, it is no surprise to find that other essential processes also occur in them. In this section we shall discuss the role of membranes in the synthesis of cellular constituents, lipids, proteins and lipoproteins with particular reference to the 'rough' endoplasmic reticulum membranes (Fig. 8.6).

Some cellular constituents are synthesised on membranes. Others, including those which later become incorporated in membrane structures, are synthesised elsewhere and moved to the membrane on or in which they function. As we have seen in Chapter 3, the lipid constituents, once synthesised, will move to bilayer formation because that is the state of lowest free energy for amphipathic molecules, but their exact positions will be influenced by other membrane molecules, particularly proteins. It is important to understand how much of the synthesis of membrane molecules, lipids and proteins is dependent on the membranes themselves.

Lipids

As we have seen in Chapter 2, amphipathic lipids are built up of units of long-chain fatty acids and hydrophilic groups by esterification with glycerol, to which other groups such as glycosides or phosphoric acid are attached. The biochemical steps in the synthesis of the saturated fatty acids from their precursor, acetyl coenzyme A (acetyl-CoA), are well understood, and the details need not concern us here. The synthesis occurs on the acyl

172

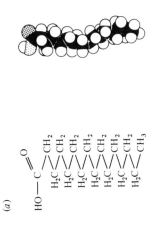

(a)

HO—C—CH$_2$—CH$_2$—CH$_2$—CH$_2$—CH$_2$—CH$_2$—CH$_2$—CH$_3$
‖
O
H$_2$C   H$_2$C   H$_2$C   H$_2$C   H$_2$C   H$_2$C   H$_2$C

Fig. 10.5. (a) Palmitic acid; (b) acetyl CoA; (c) malonyl CoA.

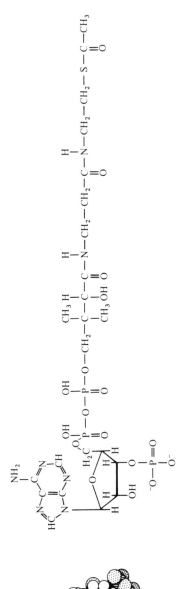

(b)

173

Fig. 10.5 (*cont.*)

(c)

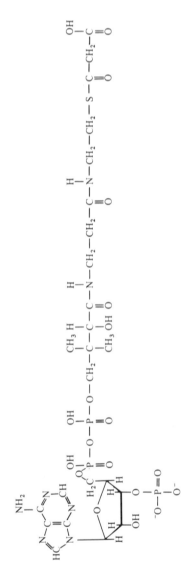

carrier protein complex in the cytosol and is not related to any membrane structure. Palmitic acid, the C_{16} saturated fatty acid formed is the precursor, in most organisms, of the other long-chain fatty acids, saturated and unsaturated (Fig. 10.5a). When first formed, it is attached to the acyl carrier protein from which it can be detached by a thioesterase, transferred to CoA or incorporated directly into phosphatidic acid. From our knowledge of membranes, we would expect palmitic acid to be attracted to the bilayer with its C_{16} tail in the hydrophobic part.

The elongation to the longer chain acids is brought about by two different enzyme systems, one occurring in the outer mitochondrial membrane and the other in the endoplasmic reticulum; both require the reducing molecule NADPH. In mitochondria, the chains are lengthened at the carbonyl end by successive additions of acetyl units from acetyl-CoA by acyl-CoA synthetase which occurs in the outer membrane (Fig. 10.5b).

In the endoplasmic reticulum (or the microsomes which can be formed from it) the elongation takes place in the membrane with malonyl-CoA as the source of the acetyl group (see Fig. 10.5c). The endoplasmic reticulum is a bilayer, with 55% ($\pm 10\%$) protein, 40% ($\pm 10\%$) lipid, 4.5% ($\pm 1.5\%$) cholesterol (Threadgold, 1976). The relatively high content of protein is undoubtedly due to the very high concentration of enzymes as well as to structural proteins. The membrane has oxidoreductases, transferases and hydrolases. For our present purposes we outline the membrane's role in biosynthesis of lipids, glycolipids and cholesterol. About 70% of the lipids in the membrane are phospholipids, predominantly phosphatidylcholine and phosphatidylethanolamine, with phosphatidylserine, phosphatidylinositol and sphingomyelin. Most of the phosphatidylethanolamine and phosphatidylserine are on the cytoplasmic side of the membrane; most of the phosphatidylinositol and sphingomyelin are on the cisternal side while phosphatidylcholine is about equally distributed on each side.

Bacterial cells, in which the hydrocarbon chains are lengthened by a mechanism similar to that described, have fatty acids with only one double bond. In higher plants and animals, as we have seen, the constituents of the lipids may have several double bonds. The desaturation, i.e. the introduction of one or more double bonds into the chains of the longer fatty acids, takes place with the acid attached to CoA, under the action of an oxidase in the endoplasmic reticulum membrane. In animal tissue, the oxidase is dependent on NADPH, cytochrome b_5 and oxygen (Fig. 10.6). In some plants and microorganisms a different oxidase, dependent on NADPH, flavoprotein, iron-sulphur proteins and oxygen, brings about the same reaction. In many bacteria an entirely different mechanism, not dependent on oxygen, is involved.

Now we have discussed how membrane-dependent reactions can lengthen the hydrocarbon chains and introduce double bonds. In making membrane lipids, the next step attaches the fatty acid chains by esterification to glycerol phosphate, and is also localised in the endoplasmic reticulum. The mechanism differs for the different lipids and we can omit the details but, in principle, the fatty acid-CoA interacts with glycerol phosphate to add the first hydrocarbon chain; then this molecule (a lysophosphatidic acid) reacts with a second acyl-CoA to give the diacyl phosphatidic acid. This can hydrolyse to diacylglycerol. The last step in the synthesis of phosphatidylethanolamine, for example, is when cytidine diphospho-ethanolamine reacts with the diacylglycerol to attach the phospho-ethanolamine and split out cytidine monophosphate. All the enzymes catalysing these reactions are tightly bound to the endoplasmic reticulum membrane. The later steps in cholesterol synthesis are dependent on squalene monoxygenase which is bound to the endoplasmic reticulum and on enzymes from the cytosol, for which phosphatidylserine of the membrane and flavine adenine dinucleotide are cofactors.

The membranes of the endoplasmic reticulum in higher organisms are thus the site of the final synthesis of the lipids. Transfer to other sites will be possible by three mechanisms. First, as we have seen, the solubility of lipids in water, though very low, is not zero and some can move through the aqueous phase to reassemble in micelles or bilayers; this however will

Fig. 10.6. The introduction of double bonds into fatty acids: (a) NADPH donates electrons to cytochrome b_5 reductase and they pass to cytochrome b_5; (b) activated by another membrane-bound enzyme, oxygen interacts with the electrons, two hydrogen atoms are taken from the fatty acid and water is formed; the fatty acid has one double bond between a pair of carbons; CoA is essential.

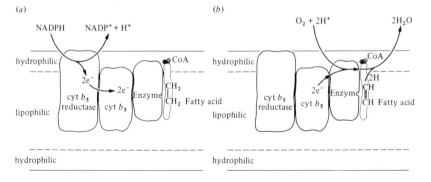

be very slow indeed. Second, there is good evidence that there are carrier proteins which take lipids from one membrane to another (summarised by Parry, 1978); this mechanism provides a much more rapid and specific method of movement. Third, as we have seen, lipids can diffuse quite rapidly along the bilayers themselves, especially in fluid bilayers. Membranous continuity between the endoplasmic reticulum and other organelle membranes has frequently been observed by electron microscopy, particularly between rough endoplasmic reticulum and smooth endoplasmic reticulum and between nuclear membrane and rough endoplasmic reticulum. Furthermore, we must always remember that these membranes are under the control of the cell and can be formed, destroyed, invaginated, reformed, pinched off and so on; in this way a variety of special mechanisms is likely to be available, including those which will transfer structural lipids from one side of a bilayer to the other.

In this section I hope I have said enough to establish that the molecules which give membranes their special properties are, themselves, dependent on membranes for their synthesis where they have the advantage of charged surfaces, hydrophilic groups, and hydrophobic layers in which to favour the different synthetic enzymatic reactions involved.

Proteins

Some membranes play an important role not only in receiving the membrane-bound enzymes at the correct places but also in their synthesis. The best understood membrane associated with protein synthesis is the rough endoplasmic reticulum, especially for synthesis of proteins being secreted by cells (Chapter 8). The polypeptides, as they are synthesised, are transferred across the membrane into the cisternal space between the endoplasmic reticulum membranes. From there they are discharged by vesicles or by Golgi bodies to the exterior of the cell. Other proteins, which go to other organelles, may also be synthesised in the rough endoplasmic reticulum and transported within the cell. See Svardal & Pryme (1980).

Ribosomes, the sites at which messenger ribonucleic acid (mRNA) interacts with transfer RNA which carries the amino acids, are responsible for the synthesis of polypeptide chains and hence of proteins. Many ribosomes are free in the cytosol or in organelles, e.g. in chloroplasts, and make proteins for the many different purposes of cells (Fig. 10.7). Some of the proteins made by the free ribosomes may be destined for membranes and will be partitioned into membranes if they are suitably amphipathic. Precursors of many of the proteins of mitochondria and chloroplasts are synthesised on cytoplasmic ribosomes. These precursors are larger than the

proteins themselves. For instance, the three largest subunits of yeast
F_1ATPase are membrane-bound on the inner side of the inner mitochondrial
membrane but their precursors enter by crossing both outer and inner
membranes and are then converted to the proteins by proteolytic action.
This change takes place in the absence of protein synthesis but is dependent
on ATP in the mitochondrion. We do not know whether the ATP is
necessary to the transfer of the precursor or for its processing to the protein
or both. Similar processing seems to be involved in some of the CF_1 subunits of chloroplasts of spinach leaf. Thus some enzymes of mitochondria
and chloroplasts, some becoming membrane-bound, are dependent on
cytoplasmic ribosomes and on a membrane transfer system which is not
understood (Nelson & Schatz, 1979; Nelson, Nelson & Schatz, 1980).
However, ribosomes associated with other protein syntheses occur between
the cristae on the inner mitochondrial membranes.

Cells which have a rough endoplasmic reticulum, i.e. with ribosomes
bound to the membrane (Fig. 10.8), are concerned in the synthesis of
proteins for secretion, proteins which will be transported to other cellular
organelles and also membrane-bound proteins, either those traversing the
membrane or placed asymmetrically on one or other side (Fig. 10.9). All
these proteins must either cross or be buried in the endoplasmic reticulum

Fig. 10.7. Protein synthesis. A ribosome free in the cell can, under the
influence of mRNA, take amino acids (AA) from transfer RNA
(tRNA) and synthesise proteins (Pr); proteins with suitable lipophilic
regions become incorporated in membranes.

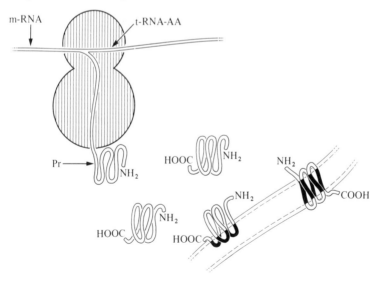

Fig. 10.8. Ribosomes attached to the membranes of the endoplasmic reticulum (ER) adjacent to a mitochondrion (M) in a mouse liver cell (electron micrograph). (Micrograph: J. M. Bain.)

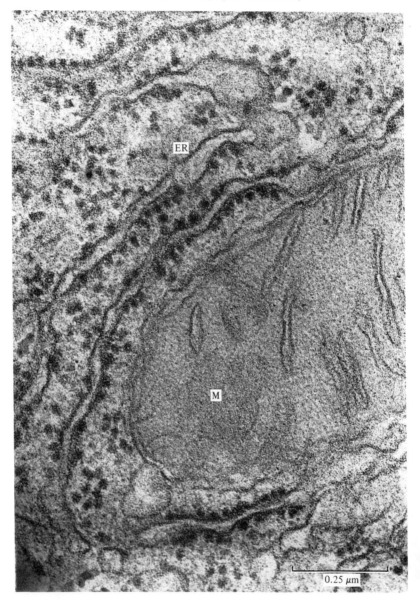

179

membrane. Many studies of secreted proteins suggest that they are synthesised by the membrane-bound *polysomes*, i.e. clusters of ribosomes connected by the messenger RNA (Fig. 10.10). Examples of secreted proteins synthesised by bound ribosomes include invertase in yeast, milk protein in the mammary gland, various hormones and egg white protein in birds.

The mechanism dependent on this polyribosome–membrane interrelation raises various questions relating to membrane properties. For example, are the polysomes attached by the subunit known as the 60S, are they attached by the polypeptide chain which is being formed or are they attached by the messenger RNA? Considerable work has been done to settle this question but it seems likely that different mechanisms may be involved in different systems. Shore & Tata (1977) conclude that binding between the 60S subunits may occur in rat, but that the growing polypeptide chains make the connection in cultured myeloma cells. In other cells it appears that the mRNA has some interaction with the membrane proteins.

Whatever the details of these mechanisms, the synthesis of membrane-bound proteins and lipoproteins within membranes seems logical. As we have seen, tightly bound membrane proteins, e.g. bacteriorhodopsin, or lipoproteins like cytochrome oxidase, have very hydrophobic portions of the molecule orientated to fit in with the hydrophobic region of the bilayer. It seems probable that the folding of the seven hydrophobic α-helices of bacteriorhodopsin would be aided by the reaction taking place in a non-aqueous environment. Similarly, the addition of lipids to proteins to form lipoproteins will be aided by the presence of a non-polar environment

Fig. 10.9. Ribosomes on the membranes of the endoplasmic reticulum synthesise proteins which may be membrane-bound.

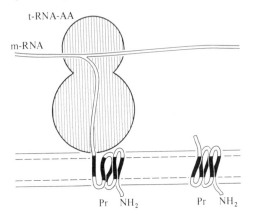

180

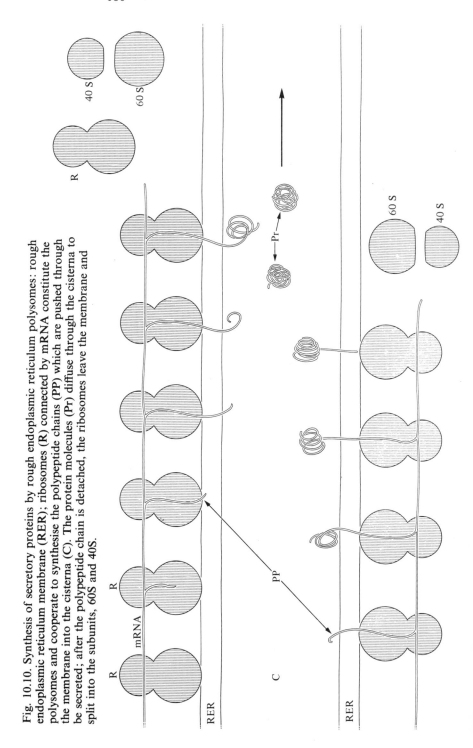

Fig. 10.10. Synthesis of secretory proteins by rough endoplasmic reticulum polysomes: rough endoplasmic reticulum membrane (RER); ribosomes (R) connected by mRNA constitute the polysomes and cooperate to synthesise the polypeptide chains (PP) which are pushed through the membrane into the cisterna (C). The protein molecules (Pr) diffuse through the cisterna to be secreted; after the polypeptide chain is detached, the ribosomes leave the membrane and split into the subunits, 60S and 40S.

for the hydrophobic forces to be set up. It seems likely that this represents an example of the way in which Nature learned, long before the organic chemist of the nineteenth century, that some reactions are distinctly favoured by the transfer from aqueous to non-aqueous environments, perhaps a very early discovery in the evolutionary process. But, unlike the method of the organic chemist, it was done on a submicroscopic, though none the less effective scale.

Suggested reading

Cuatrecasas, P. (1974). Membrane receptors. *Annual Review of Biochemistry*, **43**, 169–214.

Lehninger, A. L. (1975). *Biochemistry*, 2nd edn. New York: Worth Publishers Inc.

Lodish, H. F. & Rothman, J. E. (1979). The assembly of cell membranes. *Scientific American*, **240**, No. 1, Jan., 48–63.

Svardal, A. M. & Pryme, I. F. (1980). Aspects of the role of the endoplasmic reticulum in protein synthesis. *Subcellular Biochemistry*, **7**, 117–70.

11

Membranes and evolution

Any discussion of morphological, biochemical or physiological evolution will obviously be incomplete without an account of the lively properties of membranes and the ways they have changed in the course of time. Until recently the inherited properties of chromosomes, with their DNA carrying instructions to make particular enzyme proteins have, quite properly, been the centre of discussion of biochemical evolution but it is now possible to balance the history of enzyme development with consideration of the membranes in which so many of them act.

In this chapter I do not intend to discuss the whole subject of biochemical and physiological evolution but rather to concentrate on some places where the membrane properties we have been considering have made evolutionary advances possible. Thus, I shall consider the role of membranes in the origin of life, particularly cellular life; in the varieties of light-trapping membranes which led to both photosynthesis and eyes; in the critical reactions of membranes to temperature, both in those organisms whose temperatures vary with their surroundings and in those which have body thermostats; finally, the membranes of nerves which led to the development of brains, the physical basis of the mind of man.

The origin of life

In the 1920s, a scientist in the USSR, A. I. Oparin, and another in the UK, J. B. S. Haldane, began to speculate about the origin of life. Until that time it had been generally accepted that organic matter was always, as it is in recent times, synthesised by living activity and particularly by photosynthesis which dominates the production of organic matter now. In 1975, in Moscow, I heard Oparin, by then an old man, describe how his ideas had developed. The official translation of what he said, referring to that period, is:

182

I made bold to formulate a concept which contradicted the generally accepted opinion. According to my hypothesis, the monopoly of biogenic synthesis of organic substances is characteristic of the present epoch of the Earth's history. At the early stages of the Earth's formation, when our planet was lifeless, it was the place of abiogenic synthesis of carbonaceous compounds.

Since that original suggestion from Oparin, much evidence has been accumulated which shows that organic molecules existed, not only on this planet but also in space, as the result of chemical reactions which preceded living activity.

Radio astronomy, using the techniques of radio-frequency spectroscopy, has shown that there are many carbon compounds in interstellar space. Oró *et al.* (1978) make the important point that twelve of the monomers occurring in interstellar space can be considered as the prebiological precursors of all the biochemical compounds present in living systems. Table 11.1 (from Oró *et al.*, 1978) shows how these molecules might be related to compounds of biological importance. The conclusion is based on the parallel evidence from laboratories carrying out experiments on prebiotic organic synthesis, which have frequently used these molecules as starting points for chemical syntheses. Furthermore, the analyses of the

Table 11.1. *Biochemical monomers and properties which can be derived from interstellar molecules.* (*From Oró* et al., 1978)

	Interstellar molecules	Formulae	Biochemical monomers and properties
1	Hydrogen	H_2	Reducing agent, protonation
2	Water	H_2O	Universal solvent, hydroxylation
3	Ammonia	NH_3	Base catalysis, amination
4	Carbon monoxide	$CO(H_2)$	Hydrocarbons and fatty acids
5	Formaldehyde	CH_2O	Monosaccharides (ribose) and glycerol
6	Acetaldehyde	CH_3CHO	Deoxypentoses (deoxyribose)
7	Aldehydes (HCN)	RCHO	Amino acids
8	Thioformaldehyde	CH_2S	Cysteine and methionine
9	Hydrogen cyanide	HCN	Purines (adenine, guanine) and amino acids
10	Cyanacetylene	HC_3N	Pyrimidines (cytosine, uracil, thymine)
11	Cyanamide	H_2NCN	Polypeptides, polynucleotides and lipids
12	Phosphine (Jupiter)	PH_3	Phosphates and polyphosphates

carbonaceous meteorites which have fallen on the earth's surface have shown aliphatic and aromatic hydrocarbons, protein amino acids, non-protein amino acids and other biochemical compounds. It is believed that these were formed when the meteorite parent bodies were formed, estimated as about 4.6×10^9 years ago.

There is now a large literature on the synthesis of various organic molecules under the conditions which are thought to have prevailed on the prebiotic earth. In 1953 Miller showed that a spark in a mixture of the gases, methane, nitrogen, ammonia and water, produced both protein amino acids and non-protein amino acids, in proportions qualitatively and quantitatively like those of meteorites. Various compounds of biochemical interest have been made in laboratories using inorganic catalysts in the absence of oxygen to simulate the prebiotic conditions which were probably somewhat basic and reducing. For our interest in membranes, it is particularly important that fatty acids can be formed in a synthesis which uses carbon monoxide and hydrogen in the presence of a meteorite catalyst (Nooner et al., 1976) and that glycerol is obtained by reduction of glyceraldehyde, a product formed from the base-catalysed condensation of formaldehyde (Oró et al., 1978).

We need not discuss all the many facets of prebiotic chemical experiments and the extensive work which has been done by S. W. Fox and his associates (see Fox & Dose, 1977) on the syntheses of polyamino acids and the origins of protenoid bodies. For our purpose it is sufficient to realise that evidence points to the chemical origin of large molecules in prebiotic time.

Origin of cells

The origins of cells and the origins of bilayers are mutually dependent. As we have seen, much of the successful operation of living processes depends on the separation of one aqueous phase from another (e.g. inside and outside of cells) and also on reactions which are favoured in the non-aqueous phases (e.g. electron transport and light trapping). We do not know to what extent molecules like the lipids, and macromolecules like the proteins, accumulated in aggregates and micelles before they began to separate into bilayers. Much of the early speculation on how such molecules might be organised into the small spheres which could be regarded as primitive cells, predated our knowledge of the properties of bilayers. Oparin, for instance, was concerned particularly with boundaries or membranes formed by high molecular weight compounds. However, as we have seen in Chapter 3, the state of lowest free energy for lipid substances in water is self aggregation into micelles or vesicles, depending on the geometry of the molecules concerned. Further, other molecules, such as proteins which have hydrophobic portions, will assemble with the lipids, just as we can reconstitute vesicles today, from lecithin and a suitable protein such as bacteriorhodopsin. Given the mixtures of lipids and proteins which were capable of forming bilayer–protein membranes, vesicles of different sizes, depending on the geometry of packing, could be formed and a primitive cell structure would result. This has been discussed

Fig. 11.1. Some of the problems associated with membranes in evolution: (*a*) current speculation on the origins, in time, of organisms; dotted lines indicate uncertainties, solid lines indicate fossil evidence; (*b*) key physiological processes dependent on development of appropriate membranes.

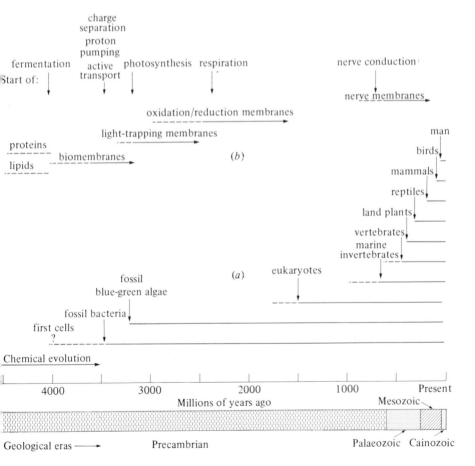

by Hargreaves & Deamer (1978). Fig. 11.1 shows some of the developments which would be expected as evolution proceeded.

Membranes start to function

Once the membranes had been formed, they would act as barriers between the outside and inside of the primitive cell. Asymmetry between the two sides of the bilayer would be expected from the beginning: first, because the curvature of the bilayer will affect the packing of the lipids and proteins in the outer and inner layers and different compositions could be expected; second, because the proteins, which are incorporated due to some lipophilicity compatible with the lipid hydrocarbon chains, and some hydrophilicity compatible with head groups in the water, are most unlikely to be symmetrical. Furthermore, anything trapped inside the membrane and changed in any way will be at a different concentration from outside. If the concentration of osmotically active molecules and ions inside the cell compared with the outside solution increased, water would enter, the cell would swell and break and only those cells which quickly achieved some mechanism for balancing the osmotic pressures would survive. Further, the primitive cell in a solution rich in organic molecules would survive only if appropriate molecules entered and waste products were eliminated.

The problems involved and hypotheses as to their solution have been discussed in detail by Raven & Smith (1982). Since any fermentation process, on which the primitive cells depended, before oxygen was free in the atmosphere, must produce internal H^+ ions and, since the entry of many weak electrolytes would require the entry of a proton with the anion, e.g. organic acids, Raven & Smith conclude that efflux of H^+ is essential if the internal pH is not to drop too low. They believe that an active H^+ transport such as is ubiquitous in extant prokaryotes was essential to the primitive cell and therefore probably evolved very early.

They suggest that, in the primitive sea where the concentration of Na^+ is believed to have been at least as high as at present, the first cells may have excluded sodium. If, then, K^+ and Cl^- could penetrate the cells, volume regulation would be possible. Consequently, they regard the high K^+/Na^+ characteristic of cells today as the result of 'an imprinting in enzyme function which was finalised in primitive, wall-less prokaryotes'. Later there became a need for an active Na^+ efflux but that was long after the primary active H^+ efflux. They also suggest that Ca^{2+} was initially excluded but later had to be actively extruded. As the metabolism of cells became dependent on phosphate at an early stage, a requirement for active phosphate influx was also met early in cell evolution. The suggestion that the H^+ active extrusion mechanism arrived early is attractive too because it is now known to energise the secondary transport (H^+ linked) of many other solutes.

The ATPases which extrude protons are not only ubiquitous but also conservative in molecular structure; mitochondrial F_1, chloroplast F_1 and bacterial F_1, which are the soluble parts of the ATPases, all consist of five subunits of molecular weights ranging from 8000 to 60000. The ATPases from different organisms are remarkably similar (Kagawa *et al.*, 1979). We can look forward to the time when complete sequencing of amino acids in both F_1 and F_0 proteins will give some guide to phylogenetic relationships.

Light-trapping membranes

There are many problems associated with speculating on the origin of the two categories of light-trapping membranes, first those like halobacteria and eyes, which have retinal chromophores and, second the chlorophyll-containing systems of the photosynthetic bacteria, algae and the higher plants. The bacteriorhodopsin system is, as we have seen, relatively simple in that it is one molecule which maintains a proton gradient. Stoeckenius (1978) has pointed out the interesting similarities of the halobacteria to the eukaryotes and their differences from the other prokaryotes. Several people have suggested that a halobacteria-like system was one of the first energy transducing mechanisms, though present halobacteria prefer respiration as an energy source and respiration is widely, though not universally, believed to exist only after oxygen, dependent on photosystem II in green organisms, was liberated to the atmosphere. How was the light-trapping phenomenon of the retinal proteins transferred to the eyes of the animal kingdom, probably much later in evolutionary time, or was it of separate origin? We do not know.

It is thought that the first kinds of photosynthesis were associated with organisms which, as the original abiotic organic molecules of the sea were depleted, were able to obtain carbon from carbon dioxide and energy from sunlight; they may have been dependent on hydrogen sulphide rather than water for the source of hydrogens, as in the purple sulphur bacteria today. Later, cells related to the blue-green algae may have been responsible for use of water and liberation of oxygen. However, if the systems resembled those of today, there would be need, not only for the light absorbing chlorophyll-type molecules, but also for a suitable charge separation and electron transport, all in a position to operate with the advantage of membrane insulation (Fig. 11.1). Similarly, the ability to fix atmospheric nitrogen would also become important, and dependent on membrane-bound enzymes. In all these systems the evolution of the porphyrins leading to iron porphyrins for electron transport and to magnesium porphyrins for light trapping, was essential.

One important hypothesis of the origin of the eukaryotes is that the free-living organisms previously developed as unicells may have entered into a symbiotic relation with large prokaryotes. The small unicells are suggested as supplying compounds which the large prokaryotes required and could no longer obtain for their own metabolism, due to the depletion of existing organic molecules. If so, the membrane structure moved in with these cells which became organelles – bacteria to become mitochondria and unicellular algae to become chloroplasts (Margulis, 1970).

Membranes and temperature

One of the great functional strengths of membranes is their ability to change easily and sensitively from quasi-liquid to quasi-solid packing of their lipids and hence influence their activities. As I have suggested (see Fig. 4.5), this property may be important at different levels in the bilayer. The centre may be more fluid than the hydrocarbon chains, on either side or on both sides, which, near the hydrophilic head groups may be quasi-solid. However, this property of being near the melting point of essential membrane components is also a great weakness. Membranes must survive and carry out their functions with fluctuating temperatures that can easily change them from one state to another. For example, a drop in temperature, but still well above the freezing point of water, could change a membrane from a more fluid to a more solid state, i.e. to below the melting point of the membrane lipids, and thus have profound effects on their physiological activities. Nor need the effect be only on the lipids; the hydrophobic forces between lipids and adjacent proteins may also be affected sufficiently to alter the rate of an enzyme reaction. Furthermore, as crystallisation of lipids proceeds, they will aggregate into domains of like molecules with exclusion of the other unlike constituents, thereby altering the normal organisation which is necessary for membrane functions.

The problem of this effect of lowering temperature will be greatest for non-parasitic microorganisms, for plants and for cold-blooded animals. It is less acute in warm-blooded animals whose body temperature is thermostatically controlled. There are now many examples of how the rates of processes in living membranes are sharply affected by temperature changes.

For many years plant physiologists, especially those who, like myself, worked on the physiology of fruits, were puzzled by the phenomenon called 'chilling injury' whereby a fruit is damaged by storage at temperatures well above 0 °C. For example, banana fruits will acquire brown streaks on the skin and deteriorate inside if stored below 12 °C, melons become pitted and rot if stored below 10 °C for some varieties and below 7 °C for others. Lyons & Raison (1970) first showed that the respiration of such cold-susceptible plants is markedly and sharply reduced at the critical temperature which corresponds to the point at which the fluid condition of some membrane lipids is decreased. Considerable work on the membrane-associated processes in chilling-susceptible plants has shown that the activities of respiratory and protein synthetic enzymes are affected at similar temperatures, but the relation to changes in lipids is not yet understood.

Many organisms can become acclimatised to new temperatures and when they do, there is considerable evidence that the proportion of saturated to unsaturated fatty acids in their membranes alters; the unsaturated fatty acids, which confer greater fluidity in the membrane, increase at lower temperature. For example, Marr & Ingraham (1962) showed that *E. coli* grown at 10 °C contained about twice as much unsaturated lipid as that grown at 40 °C. Similarly, Fulco (1972) showed that *Bacillus megaterium* increased the degree of unsaturation in membrane lipids at lower temperatures. This increase at low temperature was related to the activity of the oxygen-dependent enzyme system which puts a double bond in a phospholipid hydrocarbon chain (Chapter 10). This enzyme system was unstable and decayed more rapidly at higher temperature. A further factor which favoured saturated chains at higher temperature was a decrease in the synthesis of the oxygen-dependent system at the higher temperatures. How these organisms have achieved a mechanism for adapting their lipid nature to temperature and exactly what 'tells' them to do it (almost with an implication of intelligence, as Lands (1980) has pointed out in an interesting article) is a fascinating problem – and one of evolutionary significance.

Since there are some rather fundamental structural and functional problems involved, work which compares the effects of temperature on the membrane components of different organisms is of particular interest, especially as some animals can be warm-blooded for the warm part of the year and have their temperatures drop to zero in winter.

Raison *et al.* (1980) have compared a cold-blooded fish and a warm-blooded dog with the warm-blooded ground squirrel, whose body temperature is 37 °C when it is active in summer and falls to 0 °C when it hibernates in winter. The ground squirrel can oscillate between these two temperatures without loss of metabolic activity or heart failure. By contrast, non-hibernating mammals such as dogs, die due to ventricular fibrillation followed by cardiac arrest, at body temperatures between 20 °C and 10 °C. The authors measured the state of the molecules in extracted myocardial membranes by differential scanning calorimetry and spin-labelling ESR. In the trout, adapted to cold water, they found no sharp change in the membrane lipids over the temperature range from 40 °C to 0 °C and no impairment of cardiac function. On the other hand, the membranes extracted from a dog showed a transition, i.e. evidence of a possible shift from quasi-liquid to a quasi-solid state at 15 °C, suggesting that there could be some causal relationship with the loss of cardiac function at about that temperature. The interesting observation was that the ground squirrel that was summer-active showed some evidence of a shift in the lipid state of its myocardial membranes at about 28 °C; by contrast, the winter-hibernating squirrel showed that a similar change did not occur until the temperature had dropped to 16 °C. When the lipids were isolated from both squirrels it was found that the hibernating squirrel had more double bonds in its membranes than the summer-active squirrel, suggesting that the membranes were capable of maintaining more fluidity in the winter adaptation. These results heighten the interest in the possibility that unsaturation in membranes, to maintain fluidity, is important in their physiological activity at lower temperatures. Exactly how the reduction in fluidity at lower temperatures inhibits the activities of the cells is not known in detail and may well be different in different organisms. The problem obviously has evolutionary significance.

The brain

It is a matter of considerable interest that the human brain uses such a high proportion of the oxygen consumed by the human body. As Lehninger (1975) records this may be 20% of the oxygen consumption of the resting adult and, on a weight basis, this is more than any other organ uses. Most of this oxygen consumption is required to maintain the membrane potential characteristic of nerve cells where sodium is pumped out and potassium pumped in by the Na^+/K^+ ATPase, using the ATP formed in respiration. These resting potentials allow the brain cells to communicate with each other by action potentials in a complicated circuitry which, in a manner we do not understand, is the physical basis

191

of mind – the culmination of a remarkable evolutionary trend in the liveliest of all properties of membranes.

Suggested reading

Fox, S. W. & Dose, K. (1977). *Molecular Evolution and the Origin of Life*, revised edn. New York: Marcel Dekker Inc.
Margulis, L. (1970). *Origin of Eukaryotic Cells*. New Haven: Yale University Press.

POSTSCRIPT

In suggesting why I should wish to write a book about membranes, I said I wanted to create a picture of how lively these structures are at molecular level. Now it is written, I am even more conscious of the difficulties and of my shortcomings. For example, I have not dealt with some of the membrane phenomena, e.g. fusion, amoeboid movement etc., which would be additionally informative. I have oversimplified in many places and have not, perhaps, met the difficult task of discussing protein conformational changes as well as I should.

What will be the measure of success of this book? First, I hope it will give some (especially young) scientists an idea of the complexities and challenges of this interesting interdisciplinary field. Second, if it stimulates someone to do a critical experiment which will prove something I have said is wrong and will offer a better explanation, it will be worthwhile. Science often advances because someone 'gets mad at' what has been written and dashes off to prove it wrong. Every criticism, leading to increase in understanding, will contribute to scientific advance.

I have just begun on this fascinating theatre of the membranes. I feel like a drama critic who begins to write about a huge theatrical company with an almost endless repertoire. He can see only some of their plays, can place only his interpretations on the character parts as they unfold before his eyes but their art will continue even after he is dead. On the membrane stage, the company approaches astronomical numbers, the individual molecular actors are playing many roles, and we have not yet seen a fraction of their plays. The gradual unfolding of their many-sided art will continue to intrigue scientific theatre writers for generations.

REFERENCES

1. Lively properties – themes and variations

Anderson, D. H. & Fisher, S. K. (1976). The photoreceptors of diurnal squirrels: outer segment structure, disc shedding and protein renewal. *Journal of Ultrastructure Research*, **55**, 119–41.

Blest, A. D. (1978). The rapid synthesis and destruction of photoreceptor membrane by a dinopid spider: a daily cycle. *Proceedings of the Royal Society of London*, Series B, **200**, 463–83.

Danielli, J. F. (1936). Some properties of lipoid films in relation to the structure of the plasma membrane. *Journal of Cellular and Comparative Physiology*, **7**, 393–408.

Danielli, J. F. & Davson, H. (1935). A contribution to the theory of permeability of thin films. *Journal of Cellular and Comparative Physiology*, **5**, 495–508.

Davson, H. & Danielli, J. F. (1943). *The Permeability of Natural Membranes*. Cambridge: University Press.

Gorter, E. & Grendel, F. (1925). On bimolecular layers of lipoids on chromatocytes of blood. *Journal of Experimental Medicine*, **41**, 439–43.

Israelachvili, J. N. (1978). The packing of lipids and proteins in membranes. In *Light Transducing Membranes: Structure, Function and Evolution*, ed. D. W. Deamer, pp. 91–107. New York: Academic Press.

Singer, S. J. & Nicolson, G. L. (1972). The fluid mosaic model of the structure of cell membranes. *Science*, **175**, 720–31.

2. Composition – special molecules

Adam, N. K. (1937). Molecular forces, orientation and surface films. In *Perspectives in Biochemistry*, ed. J. Needham & D. E. Green (pp. 81–90). Cambridge: University Press.

Alfonzo, M. & Racker, E. (1979). Components and mechanism of action of ATP-driven proton pumps. *Canadian Journal of Biochemistry*, **57**, 1351–8.

194

Dickerson, R. E. (1972). The structure and history of an ancient protein. *Scientific American* (April 1972), 58–72.

Engelman, D. M., Henderson, R., McLachlan, A. D. & Wallace, B. A. (1980). Path of the polypeptide in bacteriorhodopsin. *Proceedings of the U.S. National Academy of Sciences*, **77**, 2023–7.

Gerber, G. E., Anderegg, R. J., Herlihy, W. C., Gray, C. P., Biemann, K. & Khorana, H. G. (1979). Partial primary structure of bacteriorhodopsin: sequencing methods for membrane proteins. *Proceedings of the U.S. National Academy of Sciences*, **76**, 227–31.

Hardy, W. B. (1913). The influence of chemical constitution upon interfacial tension. *Proceedings of the Royal Society of London A*, **88**, 303–13.

Henderson, R. (1977). The purple membrane from *Halobacterium halobium*. *Annual Reviews of Biophysics and Bioengineering*, **6**, 87–109.

Kagawa, Y., Sone, N., Hirata, H. & Yoshida, M. (1979). Structure and function of H^+-ATPase. *Journal of Bioenergetics and Biomembranes*, **11**, 39–78.

Langmuir, I. (1917). The constitution and fundamental properties of solids and liquids. II. Liquids. *American Chemical Society Journal*, **39**, 1848–906.

Ovchinnikov, Y. A., Abdulaev, N. G., Feigina, Y. M., Kiselev, A. V. & Lobanov, N. A. (1979). The structural basis of the functioning of bacteriorhodopsin: an overview. *Federation of European Biochemical Societies Letters*, **100**, 219–24.

Parkes, J. G. & Thompson, W. (1970). The composition of phospholipids in outer and inner mitochondrial membranes from guinea-pig liver. *Biochimica et Biophysica Acta*, **196**, 162–9.

Pockels, A. (1891). Surface tension. *Nature* (London), **43**, 437–9.

Rayleigh, J. W. S. (1899). Investigations in capillarity. *Philosophical Magazine*, **48**, 321–37.

Stoffel, W. & Schiefer, H-G. (1968). Biosynthesis and composition of phosphatides in outer and inner mitochondrial membranes. *Hoppe-Seyler's Zeitschrift für Physiologische Chemische*, **349**, 1017–26.

3. Structure – cooperative molecules

Israelachvili, J. N., Marčelja, S. & Horn, R. G. (1980). Physical principles of membrane organisation. *Quarterly Review of Biophysics*, **13**, 121–200.

4. Dynamics – moving molecules

Blasie, J. K. (1972). The location of photopigment molecules in the cross-section of frog retinal receptor disk. *Biophysical Journal*, **12**, 191–204.

Pagano, R. & Thompson, T. E. (1968). Spherical lipid bilayer membranes: electrical and isotopic studies of ion permeability. *Journal of Molecular Biology*, **38**, 41–57.

Robertson, R. N. & Thompson, T. E. (1977). The function of phospholipid polar groups in membranes. *Federation of European Biochemical Societies Letters*, **76**, 16–19.

Toyoshima, T. & Thompson, T. E. (1975*a*). Chloride flux in bilayer membranes: the electrically silent chloride flux in semi-spherical bilayers. *Biochemistry*, **14**, 1518–25.

Toyoshima, T. & Thompson, T. E. (1975*b*). Chloride flux in bilayer membranes: chloride permeability in aqueous dispersions of single-walled, bilayer vesicles. *Biochemistry*, **14**, 1525–31.

5. Energy transduction – light-energy trapping

Anderson, J. M. (1975). The molecular organization of chloroplast thylakoids. *Biochimica et Biophysica Acta*, **416**, 191–235.

Andersson, B. & Anderson, J. M. (1980). Lateral heterogeneity in the distribution of chlorophyll-protein complexes of the thylakoid membranes of spinach chloroplasts. *Biochimica et Biophysica Acta*, **593**, 427–40.

Bishop, D. G., Kenrick, J. R., Bayston, J. H., Macpherson, A. S. & Johns, S. R. (1980). Monolayer properties of chloroplast lipids. *Biochimica et Biophysica Acta*, **602**, 248–59.

Blasie, J. K. (1972). The location of photopigment molecules in the cross-section of frog retinal receptor disk. *Biophysical Journal*, **12**, 191–204.

Bonting, S. L. & Daemen, F. J. M. (1977). Structure and function of the photoreceptor membrane of vertebrate rods. In *Structure and Function of Energy-transducing Membranes*, ed. K. van Dam & B. F. van Gelder, pp. 193–208. Amsterdam: Elsevier Scientific Publishing Company.

Brigden, J. & Walker, I. D. (1976). Photoreceptor protein from the purple membrane of *Halobacterium halobium*. Molecular weight and retinal binding site. *Biochemistry*, **15**, 792–8.

Clayton, R. K. (1981). *Photosynthesis: Physical Mechanisms and Chemical Patterns*. Cambridge: Cambridge University Press.

Engelman, D. M., Henderson, R., McLachlan, A. D. & Wallace, B. A. (1980). Path of the polypeptides in bacteriorhodopsin. *Proceedings of the US National Academy of Sciences*, **77**, 2023–7.

Hubbell, W. L. & Bownds, M. D. (1979). Visual transduction in vertebrate photoreceptors. *Annual Review of Neurosciences*, **2**, 17–34.

Mitchell, P. (1961). Coupling of phosphorylation to electron and hydrogen transfer by a chemi-osmotic type of mechanism. *Nature* (London), **191**, 144–8.

Mitchell, P. (1966). Chemi-osmotic coupling in oxidative and photosynthetic phosphorylation. *Biological Reviews*, **41**, 445–602.

Mullen, E., Johnson, A. H. & Akhtar, M. (1981). The identification of Lys_{216} as the retinal binding residue in bacteriorhodopsin. *Federation of European Biochemical Societies Letters*, **130**, 187–93.

Williams, R. J. P. (1961). Possible functions of chains of catalysts. *Journal of Theoretical Biology*, **1**, 1–17.

Williams, R. J. P. (1962). Possible functions of chains of catalysts. II. *Journal of Theoretical Biology*, **3**, 209–29.

196

6. Energy transduction – oxidation–reduction energy

Berg, H. C. (1974). Dynamic properties of bacterial flagellar motors. *Nature* (London), **249**, 77–9.

Berg, H. C. (1975). Chemotaxis in bacteria. *Annual Review of Biophysics and Bioengineering*, **4**, 119–36.

Conway, E. J. & Brady, T. C. (1948). Source of the hydrogen ions in gastric juice. *Nature* (London), **162**, 456–7.

Crane, E. E. & Davies, R. E. (1948a). Chemical energy relations in gastric mucosa. *Biochemical Journal*, **43**, xlii.

Crane, E. E. & Davies, R. E. (1948b). Electric energy relations in gastric mucosae. *Biochemical Journal*, **43**, xlii.

Farrant, J. L., Robertson, R. N. & Wilkins, M. J. (1953). The mitochondrial membrane. *Nature* (London), **171**, 401–2.

Farrant, J. L., Potter, C. & Robertson, R. N. (1956). The structure of plant mitochondria. *Australian Journal of Botany*, **4**, 117–24.

Harold, F. M. (1977). Membranes and energy transduction in bacteria. *Current Topics in Bioenergetics*, **6**, 83–149.

Hinkle, P. C. & McCarty, R. E. (1978). How cells make ATP. *Scientific American*, **238** (March, 1978), 104–24.

Lugtenberg, B. (1981). Composition and function of the outer membrane of *Escherichia coli*. *Trends in Biochemical Science*, October, 1981, 262–3.

Lundegårdh, H. (1939). An electro-chemical theory of salt absorption and respiration. *Nature* (London), **143**, 203.

Lundegårdh, H. (1945). Absorption, transport and exudation of inorganic ions by the roots. *Arkiv för Botanik*, **32A**, No. 12, 1–139.

Mitchell, P. (1961). Coupling of phosphorylation to electron and hydrogen transfer by a chemi-osmotic type of mechanism. *Nature* (London), **191**, 144–8.

Palade, G. E. (1953). An electron microscope study of the mitochondrial structure. *Journal of Histochemistry and Cytochemistry*, **1**, 188–211.

Robertson, R. N. & Wilkins, M. J. (1948). Studies in the metabolism of plant cells. VII. The quantitative relation between salt accumulation and salt respiration. *Australian Journal of Scientific Research*, B, **1**, 17–37.

Robertson, R. N. & Boardman, N. K. (1975). The link between charge separation, proton movement and ATPase reactions. *Federation of European Biochemical Societies Letters*, **60**, 1–6.

Sjöstrand, F. S. (1953). Electron microscopy of mitochondria and cytoplasmic double bonds. *Nature* (London), **171**, 30–2.

7. Trans-membrane diffusion – ion carriers

Blok, M. C., Van der Neut-Kok, E. D. M., Van Deenen, L. L. M. & De Gier, J. (1975). The effect of chain length and lipid phase transitions on the selective permeability properties of liposomes. *Biochimica et Biophysica Acta*, **406**, 187–96.

197

4

Briggs, G. E., Hope, A. B. & Robertson, R. N. (1961). *Electrolytes and Plant Cells*. Oxford: Blackwell Scientific Publications.
Collander, R. & Bärlund, H. (1933). Permeabilitätasstudien an *Chara ceratophylla*, II. Die Permeabilität für Nichtelektrolyte. *Acta Botanica Fennica*, **11**, 1–14.
De Gier, J., Mandersloot, J. G. & Van Deenen, L. L. M. (1968). Lipid composition and permeability of liposomes. *Biochimica et Biophysica Acta*, **150**, 666–75.
Grunze, M. & Deuticke, B. (1974). Changes of membrane permeability due to extensive cholesterol depletion in mammalian erythrocytes. *Biochimica et Biophysica Acta*, **356**, 125–30.
Haest, C. W. M., De Gier, J., Van Es, G. A., Verkliej, A. J. & Van Deenen, L. L. M. (1972). Fragility of the permeability barrier of *E. coli. Biochimica et Biophysica Acta*, **288**, 43–53.
Haydon, D. A. & Hladky, S. B. (1972). Ion transport across thin lipid membranes: a critical discussion of mechanism in selected systems. *Quarterly Reviews of Biophysics*, **5**, 187–282.
Hope, A. B. (1971). *Ion Transport and Membranes*. London and Baltimore: Butterworths and University Park Press.
Küster, E. (1911). Über die Aufnahme von Anilinfarben in lebende Pflanzenzellen. *Jahrbücher für wissenschaftlicher Botanik*, **50**, 261–88.
Mansfield, W. W. (1958). Influence of monolayers on evaporation from water storages. I. The potential performance of monolayers of cetyl alcohol. *Australian Journal of Applied Science*, **9**, 245–54.
Overton, E. (1895). Über die osmotischen Eigenschaften der lebenden Pflanzen und Thierzelle. *Vierteljahrsschrift der Naturforschende Gesellschaft in Zurich*, **40**, 159–201.
Ruhland, W. (1912). Studien über die Aufnahme von Kolloiden durch die Plasmahaut. *Jahrbücher für wissenschaftlicher Botanik*, **51**, 376–431.
Stryer, L. (1981). *Biochemistry*, 2nd edn. San Fransisco: W. H. Freeman & Co.
Verkliej, A. J., Zwaal, R. F. A., Roelofsen, B., Comfurius, P., Kastelijn, D. & Van Deenen, L. L. M. (1973). The asymmetric distribution of phospholipids in the human red cell membrane: A combined study using phospholipases and freeze-etch electron microscopy. *Biochimica et Biophysica Acta*, **323**, 178–93.

8. Transport, absorption and secretion

Chappell, J. B. & Crofts, A. R. (1966). Ion transport and reversible changes of isolated mitochondria. In *Regulation of Metabolic Processes in Mitochondria*, ed. J. M. Tager, S. Papa, E. Quagliariello & E. C. Slater, pp. 293–316. Amsterdam: Elsevier Publishing Company.
Chappell, J. B. & Haarhoff, K. N. (1967). The penetration of mitochondrial membrane by anions and cations. In *Biochemistry of Mitochondria*, ed. E. C. Slater, Z. Kaninga & L. Wojtczak, pp. 75–91. London and New York: Academic Press.

Garrahan, P. J. & Glynn, I. M. (1966). Driving the sodium pump backwards to form adenosine triphosphate. *Nature* (London), **211**, 1414–15.

Johnson, R. G., Carlson, N. J. & Scarpa, A. (1978). ΔpH and catecholamine distribution in isolated chromaffin granules. *Journal of Biological Chemistry*, **253**, 1512–21.

Johnson, R. G. & Scarpa, A. (1979). Proton motive force and catecholamine transport in isolated chromaffin granules. *Journal of Biological Chemistry*, **254**, 3750–60.

MacLennan, D. H. & Holland, P. C. (1975). Calcium transport in sarcoplasmic reticulum. *Annual Review of Biophysics and Bioengineering*, **4**, 377–404.

MacLennan, D. H. & Campbell, K. P. (1979). Structure, function and biosynthesis of sarcoplasmic reticulum proteins. *Trends in Biochemical Science*, July 1979, 148–51.

Mitchell, P. (1961). Coupling of phosphorylation to electron and hydrogen transfer by a chemi-osmotic type of mechanism. *Nature* (London), **191**, 144–8.

Robertson, R. N. (1960). Ion transport and respiration. *Biological Reviews*, **35**, 231–64.

Roseman, S. (1972). Carbohydrate transport of sodium in bacterial cells. In *Metabolic Pathways*, 3rd edn, vol. 6, ed. D. R. Greenberg, pp. 41–88. New York: Academic Press.

Steiner, D. F., Kemmler, W., Tager, H. S. & Peterson, J. D. (1974). Proteolytic processing in the biosynthesis of insulin and other proteins. *Federation Proceedings*, **33**, 2105–15.

9. Excited membranes and signal transmission

Agnew, W. S., Levinson, S. R., Brabson, J. S. & Raftery, M. A. (1978). Purification of the tetrodotoxin-binding component associated with the voltage-sensitive sodium channel from *Electrophorus electricus* electroplax membranes. *Proceedings of the U.S. National Academy of Sciences*, **75**, 2606–10.

Armstrong, C. M. (1971). Interaction of tetraethylammonium ion derivatives with potassium channels of giant axons. *Journal of General Physiology*, **58**, 413–37.

Ashcroft, R. G., Coster, H. G. L. & Smith, J. R. (1977). Local anaesthetic benzyl alcohol increases membrane thickness. *Nature* (London), **269**, 819–20.

Bangham, A. D., Hill, M. W. & Mason, W. T. (1980). *Progress in Anaesthesia*, vol. 2. New York: Raven Press.

Benzer, T. I. & Raftery, M. A. (1972). Partial characterization of a tetrodotoxin-binding component from nerve membrane. *Proceedings of the U.S. National Academy of Sciences*, **69**, 3634–37.

Breckenridge, W. C., Gombos, G. & Morgan, I. G. (1972). The composition of adult rat brain synaptosomal plasma membranes. *Biochimica et Biophysica Acta*, **266**, 695–707.

Camejo, G., Villegas, G. M., Barnola, F. V. & Villegas, R. (1969). Characterization of two different membrane fractions isolated from the first stellar nerves of the squid, *Dosidicus gigas. Biochimica et Biophysica Acta*, **193**, 247–59.

Cotman, C. W., Blank, M. L., Moehl, A. & Synder, F. (1969). Lipid composition of synaptic plasma membranes isolated from rat brain by zonal centrifugation. *Biochemistry*, **8**, 4606–12.

Fifield, R. (1980). Liposomes: bags of biological potential. *New Scientist*, 16th October, 150–3.

Hodgkin, A. L. (1951). The ionic basis of electrical activity in nerve and muscle. *Biological Reviews*, **26**, 339–409.

Hodgkin, A. L. & Huxley, A. F. (1945). Resting and action potentials in single nerve fibres. *Journal of Physiology*, **104**, 176–95.

Hodgkin, A. L. & Huxley, A. F. (1952). A quantitative description of membrane current and its application to conduction and excitation in nerve. *Journal of Physiology*, **117**, 500–45.

Hodgkin, A. L. & Katz, B. (1949). The effect of sodium ions on the electrical activity of the giant axon of the squid. *Journal of Physiology*, **108**, 37–77.

Hodgkin, A. L. & Keynes, R. D. (1955). Active transport of cations in giant axons from *Sepia* and *Loligo. Journal of Physiology*, **128**, 28–60.

Keynes, R. D. (1972). Excitable membranes. *Nature* (London), **239**, 29–32.

Keynes, R. D. (1979). Ion channels in the nerve-cell membrane. *Scientific American*, **240**, 3, 98–107.

Mark, R. F. (1979). Sequential biochemical steps in memory formation: evidence from the use of metabolic inhibitors. In *Brain Mechanisms in Memory and Learning*, ed. M. A. B. Brazier, pp. 217–225. New York: Raven Press.

Martin, R. & Miledi, R. (1978). A structural study of the squid synapse after intra-axonal injection of calcium. *Proceedings of the Royal Society of London*, B, **201**, 317–33.

10. Membrane-bound reactions – hormones, antibodies and synthesis

Cuatrecasas, P. (1974). Membrane receptors. *Annual Review of Biochemistry*, **43**, 169–214.

Nelson, N. & Schatz, G. (1979). Energy-dependent processing of cytoplasmically made precursors to mitochondrial proteins. *Proceedings of the U.S. National Academy of Sciences*, **76**, 4365–9.

Nelson, N., Nelson, H. & Schatz, G. (1980). Biosynthesis and assembly of the proton-translocating adenosine triphosphate complex from chloroplasts. *Proceedings of the U.S. National Academy of Sciences*, **77**, 1361–4.

Parry, G. (1978). Membrane assembly and turnover. *Subcellular Biochemistry*, **5**, 261–326.

Shore, G. C. & Tata, J. R. (1977). Functions for polyribosome-membrane interactions in protein synthesis. *Biochimica et Biophysica Acta*, **472**, 197–236.

Threadgold, L. T. (1976). *The Ultrastructure of the Animal Cell*. Oxford: Pergamon Press.

200

11. Membranes and evolution

Fulco, A. J. (1972). The biosynthesis of unsaturated fatty acids by bacilli. iv, Temperature-mediated control mechanisms. *Journal of Biological Chemistry*, **247**, 3511–19.

Hargreaves, W. R. & Deamer, D. W. (1978). Origin and early evolution of bilayer membranes. In *Light Transducing Membranes: Structure, Function and Evolution*, ed. D. W. Deamer, pp. 23–59. New York and San Francisco: Academic Press.

Kagawa, Y., Sone, N., Hirata, H. & Masasuke, Y. (1979). Structure and function of H^+-ATPase. *Journal of Bioenergetics and Biomembranes*, **11**, 39–78.

Lands, W. E. M. (1980). Dialogue between membranes and their lipid-metabolizing enzymes. *Biochemical Society Transactions*, **8**, 25–7.

Lehninger, A. L. (1975). *Biochemistry*, 2nd edn. New York: Worth Publishers Inc.

Lyons, J. M. & Raison, J. K. (1970). Oxidative activity of mitochondria isolated from plant tissues sensitive and resistant to chilling injury. *Plant Physiology*, **45**, 386–9.

Marr, A. G. & Ingraham, J. L. (1962). Effect of temperature on the composition of fatty acids in *Escherichia coli*. *Journal of Bacteriology*, **84**, 1260–7.

Miller, S. L. (1953). A production of amino acids under possible primitive earth conditions. *Science*, **117**, 528–9.

Nooner, D. W., Gibert, J. M., Gelpi, E. & Oró, J. (1976). Closed system Fischer-Tropsch syntheses over meteoritic iron, iron ore and nickel-iron alloy. *Geochimica Cosmochimica Acta*, **40**, 915–24.

Oparin, A. I. (1975). The problem of life origin. Paper presented, Special session *USSR Academy of Sciences*, 250th anniversary, pp. 1–8. Moscow.

Oró, J., Sherwood, E., Eichberg, J. & Epps, D. (1978). Formation of phospholipids under primitive earth conditions and the role of membranes in prebiological evolution. In *Light Transducing Membranes: Structure, Function and Evolution*, ed. D. W. Deamer, pp. 1–21. New York and San Francisco: Academic Press.

Raison, J. K., McMurchie, E. J., Charnock, J. S. & Gibson, R. A. (1981). Differences in the thermal behaviour of myocardial membranes relative to hibernation. *Comparative Biochemistry and Physiology*, **69B**, 169–74.

Raven, J. A. & Smith, F. A. (1982). Solute transport at the plasmalemma and the early evolution of cells. *BioSystems*, **15**, 13–26.

Stoeckenius, W. (1978). Speculation about the evolution of halobacteria and of chemiosmotic mechanisms. In *Light Transducing Membranes: Structure, Function and Evolution*, ed. D. W. Deamer, pp. 127–39. New York and San Francisco: Academic Press.

INDEX

abiogenic synthesis 183
acetic acid 125, 163
acetyl CoA 171–4
acetylcholine 159–63
acetylcholine esterase 163
N-acetylneuraminic acid 170
cis-aconitate 138
action potential 150, 153–6, 158–60, 162,
   164, 190
  definition 149
active transport 134, 135, 141, 185
  definition 134
acyl chain *see* hydrocarbon chain
acyl-CoA synthetase 174
ADP (adenosine diphosphate) 36, 87, 88,
   96–8, 100, 102, 107, 109, 110, 136,
   138, 139, 142, 143, 154
ADP/ATP exchange 109, 138
adenylate cyclase 166, 167
adrenal cortex 165
adrenal medulla 144
adrenalin *see* epinephrine
agglutination 169
aggregate 50, 51, 60, 184
alanine 129
algae 100, 188
  blue-green 185, 187
amino acid 128, 144
  residues 31, 34, 35
  transport 83, 89, 144
amino alcohol 26
ammonium 126
amphipathic molecules 18, 44–50, 116–18,
   171, 176
  definition 18
amphoteric electrolytes 126
anaesthetics 54, 163

animals
  cold blooded 188–90
  warm blooded 188–90
annulus lipids 56, 119
antenna pigments 95–7
antibody 6, 165, 169–71
antigen 6, 169–71
antiport 121, 138
arabinose 141
arginine 157
arsenite 138
ATP (adenosine triphosphate) 5, 36–8, 82,
   83, 87–90, 92, 96–8, 100, 102, 103,
   107, 109, 110, 135, 136, 138–44, 154,
   166, 167, 177
ATPase 31, 33, 36–8, 87–9, 95–8, 103, 107,
   109, 135, 136, 138–40, 187
  CF_0 98
  CF_1 98, 177, 187
  F_0 36, 38, 88, 187
  F_1 31, 36, 38, 87–9, 177, 187
atractyloside 109
axon 151–63
axoplasm 153–5

Bacillus megatherium 189
bacteria 61, 188
  amino acid transport 89
  cell 5, 61, 140, 141, 174
  ion uptake 89, 140
  membrane 5, 13, 29, 30, 83–9, 101–5,
   130–4, 140
  photosynthetic 1, 82, 100
  purple sulphur 187
  respiration 5
bacteriorhodopsin 33–5, 56, 71, 77, 83–9,
   179, 185, 187